THE
FOUR SEASONS

REAL COURSE ABOUT PIGEONS

Published by

M. Joseph HEUSKIN

EXPERT IN PIGEON-BREEDING

Léopold Street, 88, FLÉMALLE-GRANDE - Belgium

Translated from French by Aug. LEMMENS

Course in four lessons, treating of the instructions about pigeon-house managing.

FIRST LESSON

Anatomy — Physiology — Kinds of Pigeons — Ideal — Conformation — Pigeon-house — The Reproducers — Exercices concerning these points.

Course, Number 4522

British Library Cataloguing-in-Publication Data
A catalogue record for this book is available from the
British Library

Pigeons

Pigeon keeping is both the art, and the science of breeding domestic pigeons - it has been practiced for over 10,000 years in almost every part of the globe. In that time, mankind has substantially altered the morphology and the behaviour of the domesticated descendants of the rock dove, to suit his needs for food, aesthetic satisfaction and entertainment. There are hundreds of breeds of domesticated pigeon arising from this common ancestor, and they are generally split into three groups; Sporting, Fancy and Utility. People who breed pigeons are commonly referred to as 'pigeon fanciers', although, as is perhaps clear, pigeon keeping is not restricted to frivolity. Flying and Sporting pigeons are kept and bred for their aerial performance, as well as for reproduction. Perhaps one of the most famous examples of these birds are the 'racing homers'; trained to participate in pigeon racing, but also extremely useful as message carriers in times of war. 'Speckled Jim'; the popular British comedy sketch in Rowan Atkinson's *Blackadder* series, plays amusing testament to their important, though not unproblematic usage in the First World War. Sporting races often have large cash prizes of anything up to one million dollars, and the *Sun City Million Dollar Pigeon Race* is perhaps the most famous of such events. There are also competitions for different types of flight, including 'Rollers' (spinning around in the air), 'Tumblers' (tumbling *backwards* during flight) and 'Tipplers' (bred for their endurance; non-stop flights of over 22 hours have been reported). Fancy pigeons, the second large grouping of pigeons, are birds which are bred not based on performance, but on appearance. Examples of fancy pigeons would include Jacobins, Fantails and Pigmy Pouters. Similar to sporting pigeons however, they are competed at shows and exhibitions - judged by expert panels on their proximity to respective breed standards. These pigeons come in all sizes, colours and types. Utility Pigeons, the last grouping of pigeons, are generally bred just for their meat, or as replacement breeding stock. The meat of pigeons is customarily referred to as 'squab', a term of Scandinavian origin, meaning 'loose, fat flesh', and is considered a delicacy in many parts of the world. Historically, squabs or pigeons have been consumed in civilisations as diverse as ancient Egypt, Rome and Medieval Europe. In the modern day, both the hobby and commercial aspects of pigeon keeping are all over the world, and various specialised societies have been founded to cater for this growing passion.

Joseph HEUSKIN

Preliminaries

*Pigeon racing is rapidly becoming more and more popu-
lar and is now practised by all classes.*

*A few years ago this sport was very little practised and
nobody would have thought it could develope as it has done.*

*In spite of difficulties it has become from some points of
view even a social necessity.*

How useful has the pigeon been during the war ?

Have not these birds also been useful to explorers ?

It is amusing and intersting, as well as instructive.

*Probably, some people like carrier pigeons for the sake
of profit; but they must read, reflect, and try to combine all
their skill to have the greatest chance of success.*

*In short, every thing must be properly done and nothing
neglected. What a great advantage you derive from this
sport, the fancier stays home, spending his time studying
instead of amusing himself in light entertainments.*

*He speaks of all his hopes etc. to his wife, his children
and his parents, and thus makes home life pleasant.*

His mind is filled with healthy noble ideas.

*The first race was organized more than one century ago.
Al that time the science of pigeons was still in darkness.*

We knew but little, about these birds the way we used to manage them was simply routine without any knowledge of pigeons, reflexion or resoning. A few races were organized in a sort of way.

Pigeons were fed anyhow, regardless of the laws of nature.

Great progress was noticeable in the knowledge of pigeons as well as in all other sciences in the last century, the century of discoveries.

Routine was abandoned and learning and science began to take its place.

Evolution continued. The sport made much progress. But in spite of this, the pigeon fancier did not want to give away all the secrets.

Indeed, some were capable of noticing certain features by observing the pigeon; some said they new the proper food for pigeons, others had special theories about pairing, moult etc., but nobody I am sure would have been capable of showing regular results, during 37 consecutive years, or could call himself professor of the science of pigeons.

Not wanting to keep those secrets; discovered through more than 40 years search work, I Jos. HEUSKIN, publish this course in order to let all true fanciers benefit by my experience and observations.

I am sure, I am the only one up to this day that has been capable of revealing those secrets. only found by myself.

Champion of the societies : « The Sparrow-hawk » and « The Justice » of Flémalle-Grande since the year 1895 ; having spent my whole life observing, and trying to learn more about pigeons.

From my very youth for whole days, I remained squat in my pigeon-house, interesting myself with the merry and instructive life of my dear pigeons, my favourites, playing and speaking with them, considering them as my true friends, learning their habits, observing their different qualities, comparing defects I found in each pigeon; selecting and improving without cease.

It is necessery to add, that such work has revealed to me many secrets.

Therefore, I can safely say on examining a subject, through my method :

1) if it is an extra-carrier ;

2) if it is an extra-reproducer ;

3) how it is paired ; (mated)

4) from which couple of eggs it comes ;

'5) the nest where it was born ;

6) the best weather for its flying the quickest.

These things that nobody dared affirm before, are the results of my scrupulous observations ; and show that my researches were throughly investigated.

Such work and such discoveries may not remain as appanage of one man only. In the interest of pigeon racing, I am giving this course in order to let all fanciers benefit by my personal experience.

Written in an easy style, so as to be understood by everybody, (for I do not want to write literature), it is divided into four lessons, each treating of a certain subject.

Our favorite sport having been a game of hazard, becomes through my method a scientific game. which is to be fostered by publishing my secrets even by those who claim to be professors.

Happy will be the fanciers who follow my advice and who will scrupulously respect the way I train their pigeons, that is to say to be faithful to What I teach them.

The drawback you formerly had, will make room for an era of learning and at the same time for an era of great pleasure for yourself and success for your pigeons.

Joseph HEUSKIN,

Rue Léopold, 88,

Flémalle-Grande

(Belgium)

Anatomy and Physiology

The study of the structure of living beings is called anatomy.

Physiology is the science treating of life and of the organic functions by means of which we live.

So, I'll speak of the physiology and the anatomy of the pigeons. It is necessary for all fanciers to know something about :

1) the general skeleton of the pigeon, and to be able to distinguish the good and bad conformation of the animal ;

2) what to do in case of sickness or wounds ;

3) the proper way to feed pigeons; calculated for the natural necessities of the system ;

4) to distinguish the organs of respiration, which is very important ;

5) the circulation of the blood ;

6) the organs of motion ;

7) the generating organs ;

8) the composition of the different senses, and in a word, all the elements combined by nature which play so great a part in the life of a pigeon, also how it behaves itself.

THE SKELETON

The skeleton is composed of :

1. The beak, lower mandibule.

2. Upper mandibule.

3. The nasal chambers.

4. The nasal bones.

5. The forehead bones.

6. The skull or brain box.

7. The eye sockets.

8. The optical nerve.

9. The cervical vertebres.

11. The lumbar vertebres.

12. The rump.

13. The tail bones.

14. The collar bone.

15. The sternum.

16. The breast bone.

17. The abdominal prolongation.

18. Coracoid.

19. The shoulder blade.

20. The humerus.

21. The cubitus.

22. The radius.

23. The wing joint.

24. The metacarpus.

25. The thumb.

26. The middle toe.

27. The inside toe.

28. The ribs.

29. Protuberance resembling the nails.

30. The haunch bone.

31. The ischium.

32. The pubic bone.

33. The hole under the pubic bone.

34. The thigh bone.

35. The fubilar (rudiment).

36. The shin bone.

37. The hind toe.

39. The inside toe.

40. The middle toe.

41. The outside toe.

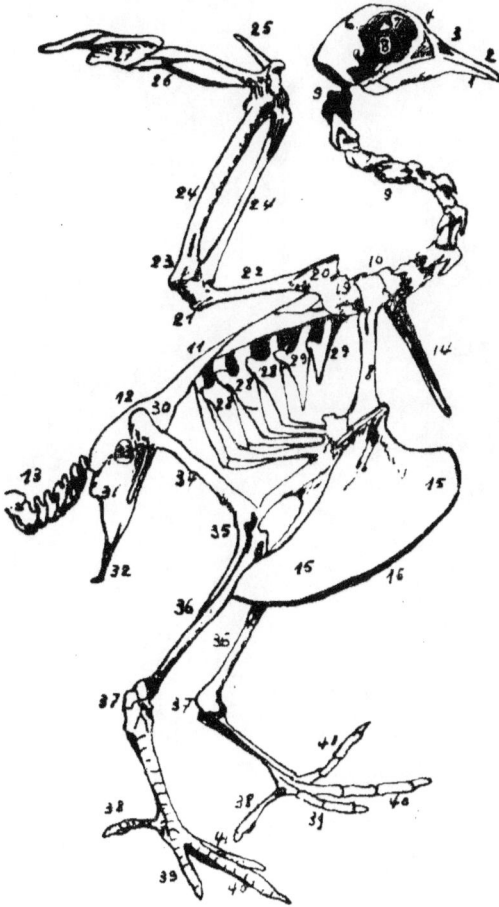

The skeleton of the pigeon.

I take the liberty to point out to you that many parts of the skeleton are not interesting to us.

I shall only speak about the most important ones .

1) the head ;
2) the neck and the spine ;
3) the thorax ;
4) the upper limbs ;
5) the lower limbs.

1. — THE HEAD.

The bones of the head are fixed together, so that nothing appears but the skull and the top of the beak, i. e. the upper maxillar and the under part of the beak, ower maxillar.

The nasal cavity, and the two cavities where the orbit of the eye is located and the two car cavities are in this bone.

2. — THE NECK.

The neck and the spine are composed of a long line of little bones beginning at the head and continuing to the tail. This skeleton supports the head as well as the different organs of the body.

There are bones in the neck called vertebres, one fixed to the other, by elastic filaments allowing great suppleness of the neck' and the movement of the head in all directions.

The dorsal vertebres continue from these, and are six in number. Each of them has two ribs, one on either side.

Next to that, we have the lumbar vertebres, as well as the different bones of the rump, these are joined together.

Then we have the tail vertebres or rump, which are very short and free, so as to let the tail move easily, which is very important during the flight when the pigeon must change the direction or the altitude.

3. — THE THORAX.

1) The thorax is composed of the sternum and the ribs. The sternum is the bone which surrounds nearly the whole body of the pigeon. It is thin, but very wide and is situated in front of the chest to protect it. Behind it we find the lungs, the heart and the liver.

In the middle of the sternum, there is a sort of crest. This crest is called the breastbone.

2) The ribs are attached to the dorsal vertebres and are fixed to the sternum. They are twelve, that is to say, six on each side.

4. — THE UPPER LIMBS.

The shoulder : it is made of three bones called : the shoulder-blade the collar-bone and the shoulder-bone protuberance.

The shoulder-bone is to be found at the upper part of the shoulder, fixed to the back above the ribs ; its shape is long and narrow.

The collar bones : the collar-bones are two, one on the left and one on the right ; they begin at the top of the shoulder-bone protuberance, at the same place as the shoulder-bones. They are thin and narrow, joining together at their extremities, to make the fork.

This fork stands in front of the chest and so protects it and the breastbone at the same time.

The shoulder-blade protu : this is attached on one side to the sternum and at the other extremity it holds up the shoulder-blade and the collar-bone.

In the cavity formed by the centre of these three bones is the head of the humerus.

The humerus : it is the most vigorous wing bone; its joints are attached the shoulder, the radius and cubitus.

The radius and *the cubitus* are in the prolongation of the humerus and are fixed to the bones of the wing joint and of the metacarpus.

The wing joint and *the metacarpus* are the bones we see at the end of the wings.

The bone of the thigh is movable, and so it allows the free movements of the metacarpus. The bones are bound one to the other and have two toes at their extremities. At the end of the radius and of the cubitus is a bone : the toe.

5. — THE LOWER LIMBS.

a) The *basin* is formed by the union of the iliac bone, the ischium and the public bone.

The public bone is free at the ends and this shows that the subject is in good health. When sound, it is strongly pressed and the extremities open when the animal suffers from any disease.

When the hen is laying, the bones open to make the way free for the egg which is at that time in the egg tube.

This is often seen about 48 hours before the hen lays the first egg.

I shall give fanciers a warning about the importance of the way in which the pubic bones must close (see instructions given in the chapter treating on the ideal conformation of a pigeon).

b) Te haunch is formed by one bone, the thigh bone ; is held up by the cavity of te basin in the iliac bone, and down by, the shin-bone and the fibular.

c) *The foot bone :* the shin-bone and the fibular are two bones forming part of the leg; the fibular is very thin and joins the other bone near the lower end of the shin-bone. It is around these bones that the most important muscles of the feet are adapted.

The shin-bone is moved up and down with the tigh-bone and by the bones of the metatarsus.

The knee-pan also stands between the shin-bone and the tigh-bone.

d) The metatarsus bone is of the same length as the shin-bone and has fewer muscles than the former. It is articulated up with the shin-bone and down with the front toes.

The part surrounding the metatarsus-bone is covered with scales; there we put the metallic ring and the racing ring.

e) The toes : the pigeon has often, I may say generally, four toes, three in front, and one behind. I have remarked during my garblings of pigeons, some with five toes and others having a membrane uniting all the toes together ; that is not a defect, for some of them were very good pigeons.

But it is not to be forgotten, that they must have one behind, in order to walk easily.

The middle-toe is composed of sensitive little phalanxes, the others have not so many, the hind one has only two.

Each of them ends in a nail, which must be a regular shape; often it is crooked through an accident when young.

THE MUSCLES

Muscles are fibrous organs formed with the flesh between the skin and the bones.

Both have at their extremities white sinews very strong and which are attached to the bones and even to the skin.

The different contractions and relaxations of these muscles produce all the movements of the limbs in which they are situated.

The well combined muscle having many muscular strings, is able to contract easily and there is no difficulty when moving these strings.

The muscle must be flexible at rest but vigorous at work. It is well known that when a muscle contracts it gets shorter, and the bones to which it is attached, also follow the same movement, but cannot become shorter, so come closer together.

Each muscle of a pigeon has its own duty, never changing.

These combined movements and the use of all these muscles at the same time help the pigeon to fly and to dart to the left or to the right, to turn, etc., etc.

THE WINGS

The main muscles of the wings are :

a) That which is attached to the crest of the breastbone and helps the wing to move forwards and upwards. It is also attached to the humerus, near the wing. It is called, the little pectoral muscle.

b) The muscle enabling the wing to go backwards and downwards is fixed to the big humerus tuberosity on one side to the sternum and to the breast-bone crest on the other side. It is called the big pectoral muscle.

c) *The different muscles stretching the wing.* — One of them is fixed to the extremity of the cubitus (near the humerus) and on the other side to the shoulder. When contracting, the cubitus changes place and causes the radius a change too and by reflexibility the muscle attached to the radius spreads out in such a way that the wing may remain quite open. The various muscles are called extensors.

d) *The different muscles shutting the wing.* — These closing the wing again are adapted to the same bones as those mentioned but on the opposite side. In this way, they move in the contrary direction, and are called the expanding muscles.

All these muscles which cause the wings to move are motor-muscles.

THE ORGANS OF DIGESTION

In speaking about the digestive organs we cannot do better than follow the course the food takes.

The food is picked up grain after grain by the beak.
It undergoes no transformation when passing through it.
The tongue within the bill is movable and pointed at the end.

We notice also at its end a number of papillas, called the gustatory papillas. They make the food palatable.

When the beak is open, we can see an opening at the upper part directly communicating with the nostrils.

This opening lets in the air to the organs of respiration; this opening is called the buccal slit (gap).

At the lower part of the bill stands an opening covered with papillas conducting the air to the lungs and which goes through the organs of respiration.

Behind this opening is a long tube; this tube is very tender, it takes the food to the crop and is called œsophagus.

On a level with the opening in the chest the œsophagus enlarges and forms the crop.

This crop is like an elastic pouch where all the food is stored; through the skin you may easily feel the grains inside.

There, food and the drink come in contact with each other.

The grains of corn become soft after a certain time; then they go in little quantities to the œsophagus where they are mixed with the gastric juice and than fall into the gizzard.

The latter is supplied with numerous little stones which by their friction one against the other act like teeth, and they facilitate the mastication of the food.

The gizzard is a very short pouch, with a thick skin, strengthened by two layers like sinews. These cause the little stones to move and thus grind the grains of corn.

After the trituration, the pap falls into the intestine, a kind of tube about one metre long (1) where it undergoes the influence of different juices or liquids secreted by the liver, the pancreas and the milt.

The Liver. — The liver which is of a reddish-brown colour is an enormous viscus, solidly constructed.

A groove divides it into two lobes almost completely surrounding the œsophagus and the gizzard.

The right lobe, where the liver vesicle is situated, secretes the bile which runs into the intestine where it helps the digestion.

The liver plays an important part in the composition of the blood, for it has an effect on the formation or the suppression of the red globules.

The pancreas is a long, narrow, white kernel, which pours a colourless liquid into the intestine.

It is the pancreas juice. This juice makes the assimilation of the grains easy.

This juice enters the intestine between the gizzard and the place where all the liver canals are connected.

The milt is an organ placed between the œsophagus and the gizzard. What it has exactly to do is not well determined, but it is known that one can live even when it is cut off.

When the food passes through the digestive tubes and has undergone the triturations, impregnations, and when it has undergone the action of the different juices, bile and

---

(1) One metre is one hundred centimetres; and a yard is 91 centimetres.

other liquids, it is transformed into excrements, which are nothing more than residue thrown off by the system.

This goes on to the cloaca from where it is ejected through the rectum; when the excrements have become liquid, or soft.

THE ORGANS OF RESPIRATION

Breathing is a regular function caused by the air going through the nose, the mouth, passing into the back of the mouth, reaching the larynx, entering the wind pipe, diffusing itself into the lungs and then into the air pouches.

The lungs are the essential breathing organs and they are placed on either side of the spine.

The change which takes place between the oxygen in the air and the carbonic acid in the blood is undergone in the lungs.

The oxygen coming in contact with the blood, becomes purified, and purifies the blood.

The bronchial tubes starting from the nostrils at the back of the mouth, meet the lungs as well as the air pouches and are connected with the cavities of the bones.

The inhaled air immediately becomes the same temperature as the body and thus reduces the weight of the pigeon.

The air pouches situated in the neck, the chest and the back, become filled, and by their dilatation they help the respiratory movements and have a great influence on the pigeon's flight.

They have also a great importance on its flying capacities: the more air they contain, the longer the pigeon can fly, and the better.

In the fourth lesson, instructions will be given on this subject.

THE CIRCULATION OF THE BLOOD

THE BLOOD.

The blood is a red liquid nourishing all the system. It passes through the whole body through the arteries and goes back to the heart by the veins which send it to the lungs where it takes oxygen again.

The blood receives all the matter coming from the digestion and it takes away the waste substances. It is composed of a liquid or serum, of red globules and of a small number of white globules.

The plasma : it is a yellowish liquid composed of about 90 % of water and 10 % of dissolved nutritive matters.

The nutritive matters contain albumen, fat, salts, etc.

The red globules : they are composed of about 65 % of water and 35 % of different substances containing principally hemoglobine the exidation of which regenerates the blood.

They are contained in the blood to the proportion of 40 % of the plasma. They give the red colour to the blood.

The white globules : these are not so numerous as the red globules (there is about one white to seven hundred or 1900 red).

The white globules contained in the blood kill the microbes.

The blood of a pigeon is bright red but changes according to the health and the conformation of the animal, the colour goes from light red to bright red.

Blood nourishes the body and takes away the waste matters.

It stops circulating only when life ends.

THE CIRCULATORY ORGANS.

The circulatory organs are composed of the heart, the arteries, the capillaries and the veins.

The heart : is a thoracic muscular organ, of a conic shape with the point downwards. It is placed behind the sternum.

It is divided into two principal parts which are the right and the left.

Each part has two compartments : the upper is the auricle; the lower is the ventricle.

The left side of the heart does not communicate with the right side. Between each auricle and each ventricle, on either side, is something like a valve preventing the blood from rushing back.

The arteries are very thick blood-vessels and take blood through the whole body. That blood is thrown out when the heart contracts.

The two important arteries are : the pulmonry artery and the aorta.

The pulmonary artery begins at the right ventricle and conducts the blood to the lungs.

The aorta starts from the left ventricle and takes the blood to the whole system.

During its passage through the arteries the blood nourishes the system.

The farther away they are from the heart, the shorter the arteries become, and the more numerous.

The capillaries are very little blood-vessels in all parts of the body, at the end of the arteries and at the beginning of veins they are so to say the continuation and the beginning of these vessels.

Being very thin they let the globules pass only one by one.

It may be affirmed that the principal phenomenons in the nutrition happen in these capillaries.

The veins : they begin at the capillaries, are not so thick as the arteries and take the blood to the right auricle (upper and lower big vein) then to the lung to be purified.

From there they take the blood back to the heart where it begins the same course again.

THE COURSE OF THE BLOOD.

After the different contractions of the heart, the blood moves from and back to the heart in the whole body of this pigeon.

The ventricules close whilst the auricles open and swell; then they in their turn close and throw the blood which they contain into the corresponding ventricles.

The blood comes back again from the lungs where it has been made pure, it flows through the pulmonary veins and is thus driven back to the left auricle, upper part of the heart.

When this is filled, it contracts and throws what it contains into the left ventricle, lower part.

The upper valve of the ventricle closes. The lower part shortens, the blood passes first of all through the pulmonary and the aorta arteries; lasthy through all the organs by means of capilaries. These vessels being so fine that it is through their very thin lining that the assimilation and the disassimilation take place after which the pure healthy blood nourishes the organ with its nutritive substances, and at the same time takes away all the waste matter, and impurities.

The blood has now accomplished its work.

From arteriel blood it has become venous. No wonder it has lost its fine red colour and is now black with impurities.

After having left the capillaries, it goes into the veins in such a condition and when passed these veins, it reaches the right auricle, upper part of the heart, where it is thrown into the right ventricle which is the lower part of the heart. The right lower part also contracts and the valve being closed, the dirty blood goes to the two lungs where it will be purified, again in the pulmonary arteries.

There, it throws off its carbonic acid and the vapour it had when in the lungs, to take in oxygen and begin again the same continual course.

If I have explained the course of the blood to you, I must now say that the contractions of the heart do not occur exactly as I have said.

The auricles, upper part of the heart contract at the same time, and the ventricles also dilate and contract together.

The lower left part, or ventricle, when contracting, beats softly, you can hear it by placing the pigeon close to the ear. That little noise comes from the heart and the number of these strokes tells you the strength of the pulse.

The beating of the pulse increases every minute when the animal becomes trained. It is there you must put the apparatus used to show the eagerness of the pigeon. This state of abundant health without which it may not be classed,provokes the temperature of the body varying from 37° to 43° or 44° centigrade.

This fact has been discovered by science.

But there was no possibility to control these variations without a special apparatus to be placed over the heart, after opening the feathers in the front part of the sternum, on the left, about 2 centimetres from the front of the breastbone, and one centimetre over the crest-whilst keeping the pigeon in the left hand.

The spiral of the apparatus is to be placed directly and entirely on the skin so as to see the changes of the quick silver in the apparatus which is held in the right hand without pressing too hard.

When there are 32° F. there is only 0° centigrade and then the 180° F. are equal to 100° centigrades.

Before each control, the apparatus must be set to indicate 35ᵈ cent. just as the doctor with his thermometer i. e. by shaking it vigorously from top to bottom.

Then look before lodging them, and during a few days, if the quick silver goes up or down.

Through these indications you can know if you may race your pigeon : for stakes only to play the little pools or the whole game.

So, you will not stake badly anymore as you will know the racing capacities of your pigeons.

Never put your pigeon in a basket when its heating powers are decreasing.

The lymphatic vessels : the lymphatic vessels are in close connection with the blood though they do not contain a single drop of it.

During the digestion they take to the liver and to the blood the chyle absorbed in the intestine.

The chyle is a liquid containing the nutritive matters of the aliments in the intestine. It is drawn from the latter through the intestinal mucous membranes.

The lymphatic system exchanges the nutritive matter for the impurities in the body.

It also takes with it the lymph which circulates through all the vessels.

THE GENITAL ORGANS

The genital organs of a cock are : the genital kernels, the outleading canals and the wand which exists only as a rudiment.

The genital kernels are two, the size of which varies.

They are situated in the lower lumbar part and in the abdominal cavity. They secrete the sperm containing the seeds which are transported through the leading canals and the wand into the egg canal, when the birds are mating.

The urine is taken to the cloaca through these passages and follows the same course.

With the coaptation of the genital organs of the cock and of the hen, both sexes are brought together and the fecundation of the eggs happens.

The genital organs of a hen are composed of the egg canal and of the ovary.

The ovary is a bunch holding the ovules or treadles which by their development form the yolk of the egg.

It is at the same place as the kernels of a cock; i. e. in the lower lumbar region.

The egg canal is a genital organ of the hen, in which the ovule when developing wraps itself with the white of the egg, and then with the shell.

The organ of copulation of a cock and of a hen is only a mere papilla.

THE SENSES

The senses are not very important from my point of view, except the sight and the finding of the cardinal points; but I prefer to speak about these things in a special chapter in order to explain it thoroughly for they are of the highest importance in pigeon racing.

With its senses a pigeon has a notion of the existence of things like all living beings.

The skin and the *feathers* let it feel objects.

The *eyes* or sight, let it see its way.

The *ears* or hearing, help it to find its way also.

The *nasal chambers* or smell are important too.

The *tongue* and the inside part of the beak gives the taste.

FORMATION OF THE PIGEON

By bringing the papillas (of the cock and the hen) into contact, the cock ejaculates its fecundating sperm.

This sperm is composed of an innumerable quantity of seeds, which progress to the genital pipes of the hen, with the help of a propelling organ, to reach the ovary.

Only two ovules take root at the same time in the ovary.

It is unusual for a hen to lay more than two eggs, for a same brood. When old and exhausted, they often lay only one egg or none.

The seeds enter into these ovules (each ovule contains only one seed), and they mingle together to form a single kernel which forms the embryon. From that time the fecundation may be sure.

Then, the ovule of the hen is to become an egg and will quickly develop itself first of all by the formation of the yolk of the egg; afterwards when it is full grown it leaves the ovary and goes into the canal of the egg, where it is kept turning round all the time.

Whilst it is turning, the yolk is being wrapped in several coats of albumen, which form the white of the egg.

As soon as the white is fixed to the yolk in a sufficient quantity, the egg is surrounded by a membrane, and it goes down farther into the canal and then it falls into a pouch· where it gets a white shell formed by calcareous materials.

The egg will be shortly laid.

The first egg is often laid between 4 or 5 in the afternoon; and the second one on the second next day between 1 p. m. and 2 p. m.

The hen and the cock eagerly begin hatching, I may say with love, sharing the work during regular intervals.

The cock often keeps on the nest from 10 a. m. till 2 p. m. and the hen does so during the rest of the time.

With admirabe assiduity, the hen hatches the eggs ; it protects them as a mother would with her children.

When the cock doesn't take its place, for some reason, (travel etc.) it only takes a very short time for eating especially during the last period of the incubation.

The hen only leaves the nest when thirsty and hungry.

Each time the cock and the hen change, the eggs are turned over; it is only after discovering this that pigeon-breeders were able to build artificial breeding machines called INCUBATORS.

The incubation lasts 18 days sometimes 17 in Summer, and in cold weather 19 or 20.

Incubation keeps the eggs at about 40° cent. and it is this temperature which makes the embryon begin its life in the eggs.

Day after day it becomes bigger and is transformed into flesh and bones.

I have noticed that a middle sized egg weighed only 23 grammes.

on the second day of incubation ... ... 22,8 grammes;

on the 3 day of incubation ... ... 22,5 grammes;

on the 5 day of incubation 22   grammes,

on the 17 day of incubation it weighed
  no more than ... ... 19,5 grammes.

(One « gramme » is the thousandth part of a « kilogr. » and the pound avoirdupois is 454 grammes and a pound-troy is 374 grammes.)

It is due to the transformation and especially to the elimination of the gas through the pores of the shell.

When looking at an egg in the light, i. e. whilst holding it between the thumb and the fore-finger, in front of a light, you may notice :

After one day's incubation, a line and a point — probably they are the head and the spine — ; after two days you see two lines on either side of the line you saw the first day; on the fourth day you may easily recognize the head, the heart and other things which are to become arteries and flesh. These all float in the liquid.

After that you will see the skin and the down beginning to form also the beak and the wings. The feet are now, whole and the lungs can be seen, they have a yellowish tint. After ten days it is difficult to recognize the different parts of the body; the whole becomes opaque.

This process is called the looming of the eggs. It can be done by means of a tube made with rolled card-board, as big as an egg and blackened inside to see more easily the different things I have just discribed.

This is only to let you know when the pigeons are laying on unproductive eggs.

After 16 days, take an egg and hold it against your ear, you will hear little beats.

Place it in the hand and it will move a little. That is a sign the youngster is living. But do not shake the shell for you might kill the bird, or injure it : a wing or a foot might be broken or twisted.

If you place an egg on an even surface, it will turn up a little in the oblique position if it is a good one.

If bad, it will remain as when placed on the surface.

The shell is broken by the youngsters, if it were broken by the parents, the shell will be broken inwardly instead of broken outwardly. You may easily see at the end of incubation period the marks which the youngster has made with its beak on the shell.

It is mechanically and unconsciously that the little bird breaks its envelope.

In case of irregular incubation on the 16th day, if you don't perceive any movement inside the egg, the youngster must be weak and it will be difficult for it to open the shell.

In a few cases you may help it : very delicately you take off the shell where you see it beginning to break and as soon as you see the beak of the young pigeon, you give it two or three drops of milk or of lukewarm wine, this very often helps, for it is always a pity to lose a youngster on which perhaps the fancier built his hopes.

After coming out of the shell, the young pigeon has yellow down, which will fall off when it begins to get its plumage. Through the heat of the parents they continue to be hatched till they are quite covered with plumage.

The youngsters have their eyes closed and something hard at the end of the beak, which helped them to break the envelope. You must never take that off. It will fall off by itself after two or three days.

During the period of incubation, you must not disturb the parents; the brood might suffer and perhaps be lost.

If a hen is annoyed by insects you have to pour some Spanish camomile powder in to the feathers as well as on the cock too.

The first food of the youngsters is the pap made by the parents.

It comes from the glands of the crop and is a composition resembling the milk of the mammiferous.

About six days they get nothing but that from their fosterers; from the 7th or 8th day they get a mixture of pieces of corn till the 10th day, but softened with digestive juices and some water, in order to make them digest. When 23 or 24 days old, the pigeon can eat by itself.

THE COMPOSITION OF THE BODY
OF THE PIGEON

A pigeon is composed, like all birds of its kind, of some water, of mineral and organic substances.

Amongst the mineral substances, the phosphoric acid and the lime are the most important in the composition of the bones.

The skeleton or frame is chiefly composed of these materials.

The sulfur helps the growth of the feathers; salt influences the digestion; the iron enriches the blood. Amongst the organic substances we notice a certain quantity of albumen, fats and hydrate of carbon. The blood, the chief maintenance of life is formed by albuminous substances combined with mineral matters to keep up sufficient energy and heat. The body of the pigeon, contains a quantity of fats. The hydrates of carbon help with a supply of starch, sugar etc.; these substances leave very little waste.

Whilst the pigeon follows the ordinary course of life, these elements become neutralized and reconstituted without cease, provided that the food consumed has been according to the needs of the body.

Therefore I shall tell you of foods which contain a sufficient quantity of the essential substances for the proper maintenance of the organism of a pigeon.

The different kinds of pigeons

THE ORIGIN OF THE BELGIAN CARRIER PIGEON

If you ask how we have obtained the Belgian carrier pigeon, you will get many different answers. But I do not think there would be one which could be proved to be exact.

What is sure is that it exists. But how has it been obtained ? From where does it come ? Or how has the cross breeding been operated ?

Though the history of pigeons is not old, many writers have written books on the origin of pigeons employed to send messages from one town to another.

If we can believe what some authors say, Persia, India, Egypt, etc. are known to have bred and tamed pigeons long before our era; the Greeks and the Italians practised it too; China had even organized a postal service and always pigeon breeding has tried to reach perfection.

About the year 1800, they began in Belgium to send messages and telegrams with pigeons. Tradition states that the Rotschild made their fortune in 1815 when they knew the result of the battle of Waterloo, for he was able to get all the rents created by the state on the London market before the defeat of Napoleon was known in England. The news had been transmitted by pigeons.

This little story told, I begin now to speak about the different kinds of pigeons.

All carrier pigeons are not good racers : they must possess certain qualities and we may say that the old race of pigeons have transmitted to our carrier pigeons some qualities and some defects.

The long flight is obtained by heredity and by training.

People always trained to get strong well formed wings to be able to effect the best flight.

The first pigeon societies began in Belgium about the year 1825 in the neighbourhood of Liege. A few races were organized without taking any notice of the direction to be taken.

Pigeons were let off anywhere, competition with no definite direction such as we see nowadays after the racing season.

Races were also organized in Paris, Frankfort and Lyons.

It is known that there was a society called « The Swallow » at Ste. Margaret, in Liege.

In these days no railway existed and pigeons were transported either on men's backs or a quicker mean, by horses.

They were shut in for many whole days, even for weeks.

The pigeons suffered mostly from bad weather, diseases or fighting and were happy when they reached Paris for they were then set free after long suffering.

If the old fanciers were living now and could see pigeons transported in aeroplanes, they would not believe their eyes.

At that time, a pigeon coming back on the same day should indeed be considered as First rate.

Under such conditions, pigeon racing would soon have been abandoned, but the invention of railways helped pigeon racing as well as industry and commerce.

Numerous societies were formed, the number of lanciers increased and the pigeons became more and more numerous.

Selection became an obligation for the sport was intersting and Belgium took a bigger interest in it than any other country.

But the science was still in a backward state, and a good selection was not effected.

The Belgian carrier pigeon was formed by involuntary cross-breeding effected in the same places, but mostly on farms.

The old types left a lot to the desired; then cross-breeding improved the pigeons. The od types from where our present day pigeons came were : The Persian carrier, The English carrier, The Fugitive-Bizet, The Tumbler, The French Cravate and The English Cravate.

THE PERSIAN CARRIER.

Description : This pigeon in its natural position, stands up straight. It is tall, with an average sized head, a long black beak covered with big slit caruncles.

The eyes are bright red and the lids very big.

Te neck is very long and rather thin, especially the part near the head.

The chest is wide and jutting out.

The body is long and is supported by long black feet. The back is also long but well curved and the wings well developed, it has a long narrow tail.

Characteristics : Traces of the Persian Carrier pigeon are to be found in our pigeons chiefly with regard to the ease with which they find the cardinal points, this is one of the most important qualities.

This pigeon was introduced first from the East into Holland and into England, from where it was brought to Belgium and to the North of France. It adapted itself perfectly to our climate, and after being transformed by the ingenuity of man, it became the English Carrier Pigeon.

THE ENGLISH CARRIER.

Description : This being the direct descendant of the Persian Carrier, the description of both are alike.

Characteristics : The English Carrier mating naturally with the common pigeons of the country has created a new pigeon called : The Dragon.

THE DRAGON.

There is an improvement in the Dragon especially in the membrane of the eyes and the caruncles of the nose. Its body is slight and well formed, the head stand out well from the neck and it possesses better racing qualities.

The beak is shorter and the neck thicker.

The chest is broader. The body becomes slighter, towards the tail which retains its original long narrow shape.

The feet are not so long, and are of a brighter red than the feet of the first pigeons of that kind.

Characteristics : This pigeon was introduced into the Brabant and Cross-bred with the other pigeons of that region. Thus improving the race. It resembles the Antwerp pigeon so much, that it is considered as its prototype.

THE FUGITIVE BIZET.

Description : The Fugitive Bizet looks very like our present carrier pigeon.

The head is round and smoothe. The beak has no big slit caruncles, the horny part is black.

The eyes are of a nice red, like cherries, with thin lids which are of a greyish white.

The wings are well formed and crossed with lines we can see in all our pigeons.

The chest is well developed, the tail is very narrow, and all the long feathers seem to have a black stripe at their extremities. This is to be seen nowadays in the case of blue and chequered pigeons. The two beam feathers which are on the outside, are white at the narrowest part except in the case of black or lead coloured pigeons.

Characteristics : The Fugitive Bizet descending from the wood-pigeon and from the wild Bizet, may be considered as being the chief origin of the carrier pigeon.

THE TUMBLER.

Description : The tumbler is a pigeon descending from the Baller, very light, with blue plumage, white throat and lapwings, bearing two or three stripes over the wings.

The head is strong, well curved and smoothe. The caruncles are very small, eyelids are thin and white.

The iris is white with a red point.

The chest is broad and well curved, the wings are wide and of an average length. The rump is completely covered.

The tail is very long and wide.

The feet have little feathers all along to the nails.

Characteristics : The name Tumbler has been given to this pigeon on account of the somersaults it effects during its flight.

Very often dangerous as well as curious because of its extraordinary contortions.

It flies very high and suddenly falls down, as if shot by a hunter, then flies up again, as if nothing happened. If any of our present day pigeons have some of these characteristics, we may be sure, they are descendants of the tumbler.

THE FRENCH CRAVATE.

There are several types of « Cravate pigeons », but the French Cravate is distinguished by the regularity of its flight. It possesses a white cravat. The plumage is all white except the shoulders which are a lovely black. They form as it were an ace of hearts, just on the middle of the back.

The head is round, the beak is short, the caruncles of an average size. The eyelids are rose coloured and the eyes are bright red.

The chest is very curved and deep, the wings are very long, the tail wide, the feet are short and the plumage abundant.

On the whole this pigeon appears to be very small.

Characteristics : The French Cravate is easily recognized from its white plumage with the two shoulders quite black. It played an important part in the creation of the Pigeon of Ghent.

THE ENGLISH CRAVATE.

Description : There are many kinds of « Cravate », but the true English Cravate is easily distinguished by its blue plumage and by the crest it has on the top of the head and also by the cravate, which has the same tint as its plumage.

It has a very big head which can be seen from behind in the middle. There is a crest which sometimes continues up to the top of the skull.

The neck is very short and thick, the cravat of the same colour as the plumage turns upwards.

The beak is short, and the caruncles are very large, but without slits or cracks; the horny part is black.

The eyes are of a bright red with a wide pupil and surrounded by a very thin whitish or greyish membrane.

The chest is very wide and deep.

This pigeon has blue plumage with two or three stripes on the back of its wings.

The wings are of average length, wide; the rump is well covered.

The tail is very narrow and not too long; the feet are short, and the plumage abundant.

On the whole, this pigeon has a very fine apparance.

Characteristics : When crossed with the « Tumbler » it produces the best results.

The Liege pigeon, sought for its wonderful beauty, has only been obtained by cross-breeding between the Tumbler and the English Cravate, and by selecting the pairs properly.

VARIOUS KINDS OF PIGEONS.

Apart from the kinds of pigeons I have spoken of, there are a few others, which are mostly considered as second rate birds, and which are not well enough known to be spoken of here. I shall only say that several of these kinds were considered only as ornaments.

But having been brought to perfection, they are no more the real types of these old races.

Not wanting to give you a description of these old races, but being determined to teach you how to breed the pigeons of to-day, I shall only say that when you possess a pigeon, having some of the characteristics of those, I have described, you may say that this bird retains a part of the qualities or of the defects found in its family;and so you will know whether you must keep them or do away with them in the progeny.

Therefore we may declare that the Belgian Carrier pigeon has been created by mating and by selection between the Persian Carrier, the English Carrier, the Dragon, the Fugitive Bizet, the Tumbler, the French and the English Cravate.

DIFFERENT KINDS OF BELGIAN PIGEONS

Our carrier pigeon was born about the middle of the last century by combining the previous types of pigeons. At that time, the race was subdivided into three classes :

1) The pigeon of Liege ;

2) The pigeon of Antwerp ;

3) The pigeon of Ghent.

To give you an idea of these different types, I shall do my best to write as complete a description as possible.

THE PIGEON OF LIEGE.

This pigeon was of average size, with a rather big head, a short beak, a large neck, eyes dark red or chesnut coloured which is much sought ofter nowadays, a middle sized pupil, surrounded by the circle of adaptation, clearly marked, the eyelids thin and of a greyish or white colour.

The neck is very short and big, the chest well formed, the feet rather short, the tail is short and narrow. This pigeon appeared to be very strong. Though it existed in all colours, they are mostly blue and checkered.

THE PIGEON OF ANTWERP.

The old type of pigeon of Antwerp had a long head and beak; the caruncles were very well developed, and the eyes of a reddish tint. They had an average pupil with a blue circle of adaptation, very big eyelids, white circle of correlation, a very long and rather thick neck, a broad chest, high feet and a long body.

The general aspect represented a very strong pigeon, recalling in its ancesters the Persian Carrier.

· Moreover it was well built with its long and powerful wings. The feathers were wide and the quill supple. The principal peculiarity was, that it had longer feet than the Liege pigeon.

THE PIGEON OF GHENT.

When speaking of the selection, I shall be able to show you, that when you cross breed pigeons to have the right sort of birds, not only things you tried to change are modified, but also other parts, such as the beak and the feet, which are transformed completely.

That is how the pigeon of Ghent sometimes transmits to its progeny characteristics like those in the two previous races.

It looks very like the pigeon of Antwerp, but there is no difficulty in recognizing each. The Ghent pigeon is much thinner and not so well developed. The bill has quite a different colour to that of the pigeon of Liege, or of the Antwerp pigeon. The latter has a black bill and the pigeon of Ghent a white one.

His eyes are of an orange colour, with a middle sized pupil and a yellow circle of adaptation ; the iris and the circle of adaptation are also yellow. The lids of the eyes are rose coloured and cover the eyes completely.

The tail is wider than that of the pigeon of Liege, and shorter than that of the Antwerp pigeon. It has an average sized beak.

The first type of this pigeon of Ghent can be distinguished by its dark plumage, It is especially noticable in the dark red, the black and the lead colour pigeons.

THE PRESENT DAY RACES

Nothing remains of those old types but the Belgian carrier pigeon, the pigeon of Antwerp, of Liege and of Ghent.

By observation and by study, we have been able to recognize the good carrier pigeons and the good reproducers inheriting from the old pigeons some qualities, which are now better formed, having such and such a plumage, the eyes showing the different circles well determined and complete. All details about these improvements will be told in the following lessons.

If nowadays we possess pigeons which can triumph over all distances, it has been due to the facilities of transport, Pigeon breeders were able to change and to buy new birds so as to produce pigeons with the necessary qualities and to have a good sort.

The old types have disappeared and made room for pigeons capable of giving good results.

Some have added new qualities suiting these already possessed by certain pigeons, and at last, all these stocks have been mixed to obtain the qualities of the Belgian carrier pigeon.

THE BELGIAN CARRIER PIGEON

We still see amongst the carrier pigeons some recalling a few characteristics of their ancestors. The neck is very short like the English cravate, we find the rose colour of the eye-lids in the French Cravate, the big caruncles and the long beak of the Persian carrier, the feathers on the feet like the Tumbler, etc.

But the real type which we find everywhere in Belgium as well as abroad, is in accordance with the transformation I have spoken of above.

In our Belgian Carrier pigeons of to-day the head is of an average size, the eyes have been greatly transformed as well as the eye-lids.

We see birds of all colours, but some have still kept a preponderant colour.

The muscles have become stronger, the size of the wings has been modified a good deal, and so has the length of the tail.

The balance is better, the instinct of finding the cardinal points is more developed.

Many of us heard years ago of the results obtained by some fanciers and one would perhaps think, that the different kinds of pigeons have been formed by them. But it is not so ! It is only by cross-breeding and selection that they have created pigeons taking the best places. It is only through perseverance that these types have become so renowned.

Before speaking of these stocks of pigeons and not of races as many fanciers call them, I shall tell you that these pigeon-breeders possessed pigeons of all colours, recalling the cross-breeds with the different kinds of the previously mentioned pigeons.

That is the reason for which I advise you to keep a well determined colour for a certain stock.

This is only with the aim of reproducing.

As far as the races are concerned, any colour will do.

There are good racing pigeons of all sorts and of all colours.

I shall try to describe as exactly as possible the most important of these first stocks, the names of which are so well known.

The Hansenne, The Grooters, The Wegge, The De Lathouwer, The Ulens and The Van-Schingen.

THE GROOTERS.

Characteristics : The ideal type of these stocks which is to be employed as reproducer must be like black velvet or motley with a yellow circle of correlation, a dark red iris, the blue pigeons having a chesnut iris, no matter what is the colour of the circle of correlation, only on condition that these birds which are to be employed as the foundation of a stock, have the circle of adaptation complete and surrounding the pupil as much as possible.

The Grooters stock is one in which we find many pigeons of value, and as excellent carriers, I shall only speak of the chequers with white feathers and of the pied having kept the natural and proper eye of that stock, that is to say, in which you may notice the different circles well determined.

The head is round and large, however it is a little longer towards the back part, than that of the Hansenne and of the Wegge.

The beak is black, of an average length, without any big caruncles; well formed and not cracked.

The forehead is very wide and high, showing a little swel- ~ling at the upper part.

The eyes well fixed in their orbits, standing very high with an average pupil.

The circle of adaptatation is often yellow.

The iris is red or dark red.

The fifth circle is visible. In the blue pigeons, the circle of adaptation is generally blue, the circle of correlation is white or blue and grey and the iris chesnut colour, no matter what may be the shade of the circle of correlation.

The eye-lids are thin and smooth of a whitish grey colour.

The neck wide and thick.

The chest very wide, curved and protruding.

The sternum is rather long and firm.

The breastbone : of an average length, neither too round nor too flat.

The wings long and supple.

The tail average length but very narrow.

The feet average length with big muscles in the thigh.

The plumage abundant and silky, the last feathers of the wing are wide and flexible.

The body is round, becoming thinner near the tail. The entire body is bulky, though it has not too much flesh. It appears to be perfectly balanced.

THE WEGGE.

Characteristics : The head which is flat on the top, is very strong, and superb.

The beak is middle sized, its horny part is black, the caruncles are not too large.

The forehead is high and broad and slightly protruding.

The eyes are well set in their orbits, with an average sized pupil.

The circle of adaptation is very thin and entire. It is of a blackish colour.

The circle of correlation is yellow.

The iris is bright red like cherries, and the fifth circle visible.

The eye-lids are white and of an average size.

The neck is very short and thick.

The chest is deep, broad and curved.

The sternum is very long and deep.

The breastbone is very long with the point a little bent.

Wings : Rather long, thick and wide.

Tail of an average length, the beam-feathers on each side are supplied with white vanes all along the outside. They resemble those of the Fugitive Bizet.

Feet rather short and very stout.

Plumage : abundant plumage, feels silky or like velvet when you touch it.

The large feathers of the wing are very wide.

A big body, a strong frame and very hard flesh.

The general aspect, is a perfect pigeon, very light, very proud, well balanced, eyes cast down a little.

In the Wegge stock, a few pigeons have a white circle of correlation with a red iris.

They are excellent racers, but they have no value with regard to reproduction.

When perfect, these birds give very good results.

THE HANSENNE.

Characteristics : The Hansenne which has given the best results everywhere and which should be chosen for reproducers, is the black chequered one and the dark one.

These pigeons are first class strugglers, never tired. They have mostly shown their capabilities when they took part in long races, in the roughest wind and during the greatest heat.

Strong head curved up to the top, front a little flat.

A thin beak of average size; the horny part is black.

The forehead high and wide, without lines between the caruncles of the nose and the top of the head.

The eyes well set in their orbits, leaving very little space between the top of the skull and the eye membrane; the pupil is average sized.

The circle of adaptation is black, sometimes orange and it completely surrounds the eye-ball.

The circle of correlation is thin and entire, of a yellowish colour.

The iris is orange or dark red.

The eye-lids are grey.

The neck is short and thick.

A very broad *chest*.

A long *sternum.*

A very long *breastbone,* with the point a little bent.

Very thick, long and wide *wings.*

An average sized *tail,* but very narrow.

Feet average length and rather red.

Very abundant *plumage,* silky, outline curved.

Body very long and back rounded.

This is a beautiful pigeon and well balanced.

Remark : In this stock, there are excellent racing pigeons of different colours which are not mentioned here.

Generally, their eyes are also different and the circles are not complete.

THE ULENS.

Characteristics : The Ulens form one of the best known stocks in Belgium and abroad.

The types which should be chosen for reproducers and which have given a large number of good youngsters are, the pale silver and the red chequered pigeons.

The head is strong, round and slightly curved.

The beak is very strong and of an average length, the horny part is black with smooth caruncles.

The forehead is wide of average height and puffed out which indicates that the bird is very intelligent.

The eyes are paced heigh in the head and are well set in their orbits, with an average sized pupil.

The circle of adaptation is entire, and of a blue colour.

The circle of correlation : in this stock, very often the pale pigeons have a white circle, and the red chequered pigeons have a yellow one.

The eyelids are of an average size and of a whitish grey colour.

Thick wide *neck* which alters and modifies the throat just under the beak.

The chest is broad, deep, and rounded.

The sternum very long.

The breastbone rather long and protuding.

Rather long *wings*, wide, thick and curved.

Tail average length but narrow.

Feet rather short and strong.

Plumage rich and abundant.

The whole body has the shape of a pear and is rather long.

The back is very curved.

Its general aspect is a well balanced and perfect pigeon.

Remark : In this stock, we have pigeons of all shades, chiefy blue and light coloured, as well as chequered pigeons.

They are excellent racing pigeons especially in a North wind.

But it has often been noticed that they do not possess the qualities required for reproducing.

THE DE LATHOUWER.

Remark about this stock : The de Lathouwer is a first class struggler, and is never tired. It can do without food for a long time. It wins the best places in the long races, chiefly in a North wind.

The pigeons of this stock which should be chosen for reproducers are the pale chequered ones. There are also some blue or pied pigeons in this stock.

The latter are not good reproducers but they come first in the races, ever if the weather is very bad.

Characteristics : The head is pretty, strong, and rather flat on the top.

The beak is of average length, very wide, with smooth nasal caruncles.

The horny part of the beak is black.

The forehead is very high and broad without lines between the caruncles and the top of the head.

The eyes are well set in their orbits, never cast downwards.

The pupil is average sized.

The circle of adaptation is blue black and surrounds the pupil completely.

Te circle of correlation is thin, entire, and of a yellowish shade.

The iris is large, and orange or bright red.

The eyelids are average sized, white or white and grey.

The neck is very thick and appears rather short.

The chest is wide and protruding.

The sternum is long.

The breastbone is very long and can be easily felt.

The wings are very long, strong, and thick.

Tail : the width of one feather is the average width.

Very strong *feet,* average length.

Plumage rich and abundant.

The *body* has the shape of a pear and is long.

The back is very curved.

The general *appearance* is perfect.

A better pigeon is not to be found.

THE VAN SCHINGEN.

Description : The Van Schingen, which resembles the Ulens is a beautiful pigeon.

The rump is quite covered with feathers. They are easily distinguished from the other birds I have described, for at first sight they give the impression that they are always sitting.

They always look straight in front of them, never casting the eyes downwards.

The red chequered, the pale and the silver coloured pigeons should be chosen for reproducers.

The birds of this stock being of different colours to those just mentioned, may be considered as good racing pigeons.

Characteristics : Extraordinary fine *head*, well rounded.

The beak is of average length, rather wide; the horny part is black.

The caruncles are smooth.

The forehead is broad, high, puffed out at the upper part which shows intelligence.

The eyes are well set in their orbits and are high up in the head.

The pupil is average sized.

The circle of adaptation is surrounding almost entirely the eyeball.

It is grey or blue.

The circle of correlation is thin, entire and white.

The iris is of a greyish or orange tint and can be seen easily.

The eyelids are thin, white or white and grey.

The neck is very short and thick.

The chest is wide and well formed.

The sternum is large and long.

The breastbone is long.

The wings are very long, wide and thick.

The tail is average length and very narrow.

The feet are very red and short.

The body is rounded and diminishing in form like a pear towards the tail.

The general aspect is of a perfect pigeon and of unequalled beauty.

GENERAL REMARKS

Now I have done my best to describe to you the general aspect of good pigeons, so that you can recognize them easily without the help of others.

The race is subdivided into three different classes, which by their evolution have created stocks quite different one from the other.

The dovecot of these well known breeders were no doubt composed of the best racing pigeons and the finest birds existing at that time.

However in spite of their fame, from the great number of pigeons I observe every year, I must say that I don't see any more specimens of these kinds.

I sometimes see pigeons having certain characteristics of these races. But I never have the plasure of examining any of the pure breeds, for these disappear rapidly.

It seems useless to tell you that every year I select thousands of pigeons in Belgium as well as abroad.

Therefore I must warn you against advertisements which say that one is selling the best stock. Be very careful, for often you may pay a high price for a pigeon which is worth nothing.

I think, that nowadays it has become impossible to have pure blooded pigeons of one of these stocks, on account of crossbreeding, when the sport became so important and widespread.

However, we have now very good birds which are sometimes even better than the first pigeons; every year, we have the opportunity of noticing this.

1) First of all, what is the reason of this ?

Because pigeons are now found everywhere.

2) Real fanciers have tried every means to improve their stocks, and so they helped the evolution of the pigeon.

3) The crossbreeding between the different pigeons have combined the great qualities of racers and reproducers.

4) The long races, the Northwinds, the warmth, etc., have eliminated the pigeons which could not endure the hardships.

5) The selection was not altogether made by the fancier; only good pigeons are kept, which have flown in good weather.

All these things enable me to say, that the best pigeons of to day are superior to those of the old stocks.

With time this sport will become more developed.

In the next lesson I shall speak of the ideal conformation of a carrier pigeon ; I shall describe the different characteristics of a good pigeon so that you will know a good pigeon from a bad one, and if it is a good racer or a good reproducer.

In this lesson will be given full particulars and knowledge about the eyes ; so you will have no doubt as to what has been previously said.

You will also be able to judge whether your pigeon can produce good youngsters or if it would be better not to let it breed.

After having studied these points you will not mind anymore from what stock your pigeons have come.

All my instructions should be carried out to the letter, and no alterations should be made.

To enable everybody who reads « The Four Seassons » to distinguish a good bird from a bad one, is my chief aim. It is the most important point of all ; and the fancier who wants to succeed, must know that.

I should like to help you in your choice, and not let you buy pigeons of no value, which you see advertised.

In every dovecot there are good birds; but there are also pigeons that only eat up the grain.

Unhappily, the fancier likes very much to keep all his pigeons, and does not always think, that to become a good breeder, he must only keep first rate pigeons.

I think it useless to repeat, that you must not buy expensive birds said to be of a good stock; only a few pigeons of a stock are worth anything. When creating a new loft the fancier must carry out all my instructions carefully.

The Ideal Conformation

In writing this book, I think it my duty, to teach new and old fanciers, what they should know, when they have but little time and little money, to set up a colony, capable of becoming famous, and how to choose amidst several pigeons the one which can improve the stock.

I have already said, that it is not possible any more to find a pigeon with the characteristics of the old races, either the pigeon of Liege or the pigeon of Antwerp or the Ghent pigeon, and I dare affirm that not a trace of the pure blood from which we got our Hansenne, Grooters, etc., exists.

Since the beginning of the XXth century, experts have done their best to keep the chief qualities of certain breeds, in their stocks, but at that time, they did not know any better.

In order to have good pigeons only, the fanciers chose a bird in the loft of a neighbour or a friend which they thought capable of improving their colony.

Others, to keep a type of pigeons which were ideal for them practised consanguinity which is really the best way.

But there is danger if you do not know exactly what to do.

Crossbreeding and selection have caused the old types to disappear.

And thus our present day pigeons remain, which possess more and better racing qualities.

These pigeons are without a doubt better than the old sort, but to judge a pigeon properly, it is necessary to know its physical conformation.

PHYSICAL CONFORMATION

1) **Propelling organs** : To be of the ideal conformation, the pigeon must be well balanced, having good moving power and a good body frame. The best conformation is, that which allows the pigeon to fly a long distance without getting too tired.

For this purpose, it must have large strong muscles.

a) The chief muscles are : **the pectoral muscles.** I shall explain the large pectoral and the little pectoral muscle.

1) The large pectoral muscle is at the extreme end of the sternum and continues to the lower end of the breast-bone, passes the shoulder joint, and joins the big joint of the humerus.

2) The little pectoral follows the same course as the large pectoral ; but it lodges just under it in the hollow formed by the sternum and the crest of the breastbone.

It is joined to the humerus and turns round the shoulder-blade protuberance.

They lean over the breast-bone, which must be very strong.

To see if a pigeon is well built, you take one in your hand.

You rub your finger along the shoulder and the breast-bone.

You feel if the muscles are large, very hard and resistant.

THE WING

b) The ideal wing should be covered on the upper part with different kinds of quills, called :

1) Primary quills ;
2) Secondary quills ;
3) The little feathers forming the mantle.

THE PRIMARY QUILLS

The primary quills are set in the centre of the wing. They are protected by the upper and the lower mantle. The quills are commonly called lapwings and play a great part in the flight.

When looking closely at them, you can perceive very interesting points. They must be as follows :

These of the cock should be very big and tapered at the ends ;

These of the shoulder be more tapered and narrower.

The centre part of these quills is called « stalk » ; it must be as thin and as supple as possible.

It must have no stripes on it. By this stalk you will see if your pigeon suffers from any disease.

In such a case, the stalks are covered with lice.

In order to see them, you open the wing of the pigeon, and examine it in a well lighted place. Sometimes you can see about fifty, all on one of these quills. They stick closely one to the other. (See the picture).

These insects are injurious to the health of the birds, and prevent it from being classified in the races.

Well formed quills must not be wavey. Such feathers show that the pigeon has been either suffering, or has had improper food.

It is a very good thing to observe these quills during the resting season.

I shall speak on this subject again in the fourth lesson.

Split quills are always bad.

Do not confound a split quill, and one that is not consolidated, or joined properly.

It is split, if there is a crack on the upper part of the stalk.

To see this crack well you must hold the pigeon before the light.

A bird having cracked quills is of no more use, and must not be kept either as a reproducer, or as a racer.

Those with black or impure blood are also to be destroyed.

The barrel, which is the black part of the quill, is fixed into the flesh.

A healthy pigeon that has properly moulted, often has the 8th., 9th., and 10th. quills of the same length.

These three feathers overlapping one another, allow only the 8th to be seen.

They extend to about 1 1/2 cm. from the end of the tail, without crossing over one another. (The inch is equal to 2.5 cm.).

Pigeons whose wings are overlapping at the extremities, are declining, and this should be noted when pairing reproducers.

The quills of ideal conformation should all form an even line at the ends, and are set horizontaly on the upper part of the tail.

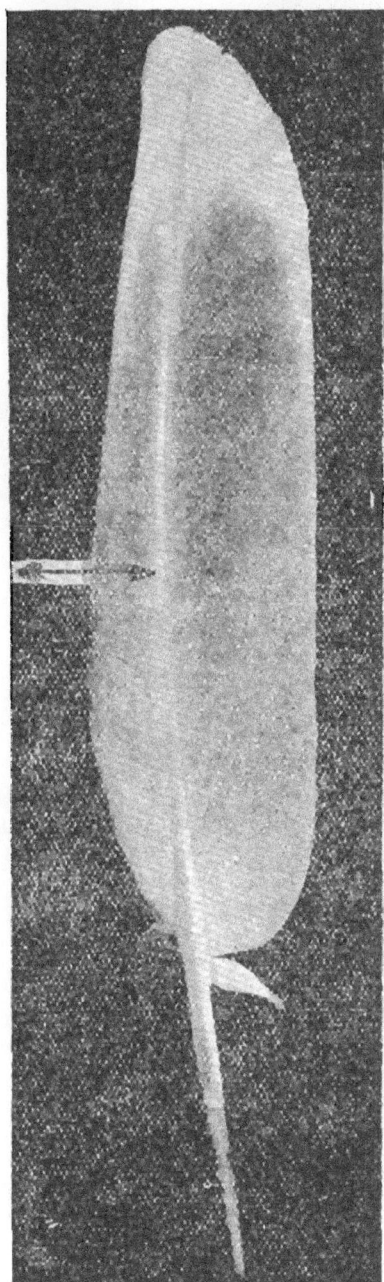

When the wing is open, the primary quills form one straight line with the secondary ones.

THE SECONDARY QUILLS

The secondary quills are feathers forming what is called the rear of the wing. They are generally twelve in the front part of the wing.

Some fanciers say, that the age of the pigeon can be told by them. But this is not certain.

When the secondary quills are of an ideal conformation, they should form one straight line with the primary quills when the wing is open.

The first secondary quill which is the nearest to the body, must be shorter than the twelfth.

The twelve secondary quills are of the greatest importance.

The pigeon will be able to cover a greater surface and to fly more quickly and easily over a long distance, when the secondary quills are short and wide, and the primary quills long.

The rear part of the wing is composed of short wide feathers, bearing a striking resemblance to the wing of the swallow. It is supported by the lower mantle which should be rich and abundant.

It is only when the pigeon has got all its primary feathers that it begins to lose those in the rear part of the wing, for it would not be possible for it to fly if all the quills fell off at the same time.

THE LITTLE FEATHERS IN THE WING

These feathers form the mantle of the wing. They also play a great part in the general aspect of the pigeon.

The upper mantles must be soft like velvet and abundant.

The better the shape of a pigeon is, the greather its value.

The lower mantles should completely cover the lower part of the wing. They should be very wide and long in order to support the bird in the air and help it in its flight.

THE STRUCTURE OF THE WING

Well formed wings must have strong, large and powermul muscles.

The pigeon can fly for several hours and cover very long distances without fatigue.

The wings should be very thick, supple and soft to the touch.

The « shoulders » or top part of the wing should be very wide and slightly curved, descending to the front end of the breast-bone when the pigeon is resting. They are hidden by a quantity of short feathers, slightly curved at their extremities.

When they are strong, the pigeon will keep flying for a long time without getting tired. They work like oars in the air.

, THE TAIL

The tail is divided into two parts, each of six quills, called tail-feathers.

I have seen pigeons with 14 and even 18 quills. But it was rather a defect than a quality ?

These tail feathers are covered with thick silky down.

The upper mantle begins at the rump and continues to about 5 centimetres from the end of the wing. (5 ctm. = 2 inches.)

It protects the quills from all the inclemencies of the weather.

The lower mantle is under the tail feathers, and is also divided into two parts consisting each of six feathers. It is called the supporting mantle. It continues to within two or three centimetres from the end of the tail. It prevents the air from passing through the tail and thus helps the pigeon in its flight.

This mantle must be wide at the beginnig and diminishing to a point near the tail.

A **long** tail denotes a bird that will fly well in a **South** wind, and a **short** tail denotes that the pigeon will fly well in a North wind.

The ideal tail is **middle sized.** Pigeons having such a tail will be good racers in all weathers.

THE BREAST-BONE

The breast-bone constitutes the point to which the pectoral muscles are attached.

The ideal breast-bone is very strong and hard. Its length varies from 7 to 9 centimètres. From the length of the breast-bone you can know if your pigeon will be fit for long, average or short distances. I shall speak of this again in the fourth lesson, when giving certain particulars and secrets.

It should be slightly curved at the breast and must be on a level with the joint of the front part of the wing, (called the shoulder by fanciers).

It should be neither too rounded nor too flat.

The sharper the breast-bone is, the greater the resistance of the pigeon will be.

When well formed, the beginning is pointed and slightly curved in front and ends at the fork to which it seems to be attached.

The fork especially **of a cock** should be short and not too thick.

That of a hen can sometimes be two centimetres long.

As I have already said, the breast-bone must be very strong to hold the pectoral muscles and to help the flying powers.

THE RUMP

The rump is at the extremity (upper part) of the tail vertebres. By it, you can see if the pigeon is healthy and not suffering from any disease.

It is easy to see, and important to note.

If it is a violet red, and letting out a clear white liquid, or sometimes yellowish ; your pigeon is suffering from internal inflammation.

If it is white and hard, it shows weakness.

When the bird is healthy, the rump is rosy.

The youngsters, whose first quills don't fall easily, very often suffer from inflammation.

As soon as you see this, open the tail which is composed as I said of twelve feathers divided into two parts of six feathers each.

Pull out two quills in the middle of the tail, one feather on each side.

Those feathers are the first two which should fall off.

Then give them some quitchgrass (triticum repens) tea or infusion. It is the best for inflammation.

One soup-spoonful to a litre of boiling water, and give to the pigeons to drink when it is cold. (one pint = 0.567 litre.)

THE ELBOW JOINT

An excellent quality in a pigeon is a strong elbow joint. It must not extend over the back.

When this joint is close to the body, the bird is a good racer in a North Wind ; when far from it, it flies well only when the wind is favourable.

But in a well formed bird, this joint is just a short distance from the body.

THE BODY

A pigeon that is well shaped or constructed gives the impression when taken in the hand, that it is composed only of wings, feet and tail.

Perfect pigeon.

— 83 —

All the weight is in front.

The flesh on either side of the breast-bone should be hard.

The chest of either cock or hen should be curved, broad, and well developed.

A broad chest shows that the pigeon has large air pouches and good muscles, without which it cannot fly during a long time.

The good pigeon will stretch out its wings when taken in the hand ; this shows that it is a pigeon for long races and not for speed flights.

To feel the resistance of the wing, try to open it when it is closed down.

The ideal body looks like an egg in shape, round but diminishing something like a pear near the tail.

The back should be curved, wide and puffed out.

A pigeon with a flat back should be destroyed, for when it is raining of in the fog, it becomes too heavy to fly well.

THE BALANCE OF A PIGEON

The pigeon of ideal conformation must be well balanced.

It seems to be impossible to be able to see if a pigeon is well balanced or not; but put it into a basket and let it rest.

You will notice after a time if : —

1) *The height is equal to the length.*

That is to say, that from the breast-bone to the end of the wings, it must measure the same length as from the bottom of the feet to the top of the head.

2) *The breadth and the height must be equal.*

This means, that the width of the forearms (called « shoulders » by some fanciers), is the same as the height from the front end of the breast-bone to the top of the head.

3) Both wings, when looking at them from the front, must be of the same depth, drooping down as low as possible, so as to be on a level with the front end of the breast-bone.

They will form a horizontal line, without overlapping at their extremities.

The outside part of the long quills of the wings comes just over the upper part of the tail.

4) The forearms should be of the same thickness and curved.

The Auxiliary Organs

THE ORGANS OF RESPIRATION

The organs of respiration are chiefly composed of the lungs and the air pouches.

The oxygen in the air purifies the blood passing in through the lungs as I have alrealy said, in page 27.

The air pouches must be well developed to make breathing easy, and to allow the pigeon to fly very fast and for a long time.

It can fly during hours and even for a whole day without opening its bill to breathe in bad weather.

When a pigeon has well developed air pouches, the chest and the back feel like elastic to the touch.

Remark : These pouches do not exist when the pigeon dies.

THE BEAK

To be able to fly a long time, both mandibles of the beak must close quite well one over the other.

If the pigeon keeps its bill open while flying, it is not a good pigeon for racing nor for reproduction.

It shows, that the pectoral muscles are weak and that the bird will readily contract infectious diseases.

The buccal opening must not be closed or stuck together; it is a sign that the bronchial tubes are obstructed.

The beak must not be too large, and the caruncles smooth.

Generally an old pigeon has larger caruncles but they must always be very smooth.

A subject with cracked caruncles will not easily find its way.

It often happens, that a pigeon is a very good racer for the first two or three years, and then cannot find its way any longer. This is because the caruncles are cracked.

When the mandibles do not overlap each other properly, the pigeon will never follow the others in a North wind.

When the upper mandible is larger than the lower one, it is very bad. For instance when eating, the pigeon pushes the grains about, instead of picking them up. Put the food in an earthenware bin, a thick layer so that it can eat at its ease.

Such a pigeon if not carefully looked after, wil never be a good reproducer, and will become weak, for, when in a basket, it will not eat enough.

Do no let your youngsters be fed by parents having big caruncles.

The beak of the young pigeon being still tender will be distorted.

The buccal split inside of the beak should be wide open, so as to let in the air easily to the lungs.

This is a sign of good health.

If this split is closed, give the pigeon honey and the bronchial tubes will get cleared again.

Put a teaspoonful of pure honey into half a cup of hot water ; clean the drinking vessel and pour your mixture into a litre of water. (a pint = 0.567 litre.)

Do this twice a weak during the racing season, and I advise you also to do it on the day of the race or on the day after coming back.

A long beak shows a good subject for a long or half length races.

A short beak shows a pigeon suitable for short races. With a middle sized beak, the pigeon will be good for any race.

When speaking of mating and classing, I shall give further particulars about the bill.

THE TONGUE

The tongue is attached to the gizzard and to the lower part of the beak. It is a movable organ and very important with regard to the swallowing of the food. It is composed of a long frame which makes it move easily. It is covered with gustatory papillas.

With the tongue the pigeon tastes what it eats.

The upper part reaches the gizzard and resembles a narrow gutter ; one centimetre long, four or five millimetres wide, ending in a point.

You can see if a bird is in good health by looking at the tip of the tongue. If it is white, it is a sign that the pigeon is ill.

If it is black, the pigeon is suffering from anemia or has impure blood ; sometimes caused by overwork, or that it has been kept too long in a basket, etc.

But do not mistake this for a blue tongue. The pigeon with a blue tongue is too vigorous.

This can be noticed in good racers and pigeons of great speed.

The tongue of a healthy bird is rosy.

THE NECK

To be ideal it must be short, thick and strong. When it is too long, the muscles are not very strong and often the pigeon is not a good racer.

A pigeon with too thin a neck must be destroyed. It has not good muscles and will not follow the other pigeons when coming back from a race. Do not forget that the neck is as important as the feet and the wings. So it is the reason for which it must be strong.

When flying during hours, sometimes for a whole day, the neck gets as tired as the feet.

If you watch a pigeon when racing, you will see that, it keeps turning its head from left to right, all the time.

The neck joints as well as the structure must be strong.

A strong animal or a strong man has a stout neck.

THE FEET

The feet should be in proportion to the whole body that is to say middle sized.

Good big muscles in the thigh, becoming gradually smaller near the knee.

The part between the knee and the toe nails should be of a rather bright red, thick and smooth.

The feet, as I told you, are very important.

The pigeon keeps them in a horizontal position while flying.

I insist upon the fact that a pigeon ought to have strong feet.

If it flies down, you may be sure that it is because its feet are tired.

Very often its failure to return from a race is due to bad feet.

If the feet become too thick and the ring gets too tight, put a drop of oil upon it. Turn the ring round and take off the few scales which surrounded it.

It is a very good thing to do, and should be done often.

THE PLUMAGE

The plumage should be soft and flexible, silky and abundant.

When you take a pigeon in your hands, the plumage should have a soft feel like velvet. If closed up in an aviary, it wil never have silky flexible plumage like the others, that are allowed to fly out of doors. But it can be a good reproducer.

An important remark is to be made here.

Never keep an excellent reproducer in an aviary more than one year.

After that time it loses its capability of finding the cardinal points, and is not energetic enough. The pectoral muscles become rounded, the air pouches swollen etc.

Experience has given the following results.

When a pigeon, that is let fly every day in the open air has colourless plumage, it is the sign that :

Its health is not good ;

It has eaten improper food, or it has endured hardships during the race; or the moult was regular, or it did not receive proper treatment during the different seasons.

THE IDEAL HEAD

It must be middle sized and rather flat on the top. Then it continues in a graceful curve to the neck.

The forehead must be broad, high and fluffy. It descends gradually to the morils and should have no wrinkles.

The eyes should be wide and placed very high in the head, as near as possible to the top. The pigeon should always look straight in front, the eyes cast slightly downwards. The pupil should be situated in the prolongation of the split of the beak.

This is to be seen in all good pigeons.

The circles of the eyes must be well marked.

A pigeon of value has often a bigger eye than a common pigeon.

It watches you wherever you go, for it is very inquisitive.

The membranes are white or grey and encircle the eye almost completely. A pigeon with large membranes flies quickly.

In a fast racer they appear oval.

The ideal pigeon has thin membranes entirely incircling the eye.

The beak is, as I said before, of an average size and smooth.

The hen of great value genarally has two little lappets under the lower mandible, one on either side.

The neck is short, very strong and thick.

The ears ought always to be covered with short feathers, without which it is not a perfect bird for it will make noise in the dovecot or in a basket. Often it is wild and does not accustom itself easily to the others.

It is very suspicious and often cries « Hoo, hoo ».

Take it out of the dovecot when it comes back from a race, for if it hears the slightest noise, it will cry out and frighten the other pigeons and prevent them from landing.

When everything is quiet in the dovecot, if suddenly the birds hear « Hoo, hoo », they will all become firghtened.

Such a defect in a pigeon is very difficult to cure.

Often I heard that noise, and I did my best to prevent it in future ; but I was obliged to kill the pigeon ; there was nothing else to be done. I tried punishing it. But to no result ; it would continue.

EXTRA-REPRODUCERS

All pigeons of the ideal conformation are to be considered as extra reproducers. You have to do exactly what I told you without changing a single thing.

You must mate birds being as much alike as possible. But do not forget that the hen is often not so big as the cock.

They must also have perfect eyes. I shall speak of the eyes in a special chapter.

To breed under good conditions, I advise you to try, to have as many couples of breeders as of reproducers.

In this way, you can rear four couples of youngsters every year, with a couple of reproducers.

As soon as they keep the nest and that the hen has laid its first eggs, they hatch them and rear the youngsters.

Soon after that they lay a second couple of eggs that you give to the breeders specially prepared for such work.

Then you let the parents hatch and rear the first youngsters, but you must take away the cock and the hen alternatively for a whole day.

When the youngsters are old enough, you separate the cock and the hen during two or three weeks. In doing so, they do not get too tired, and you can thus obtain two first couples of youngsters from pigeons which are not tired, and during the same year, four other couples of youngsters from each pair of reproducers.

During the time they are separated, they must be allowed to fly about one hour, and this at least once a day in their turns. If they are with other pigeons, let all be of the same sex.

Do not allow a reproducer to breed after the 15 th of August.

At that period of the year, the fine days become rare and begin to shorten; the sun is not so warm as earlier in Summer and the youngsters do not grow so well as the Spring birds.

Do not think that you will become famous by breeding a large number of pigeons. Remember that the proverb says, « Quality is better than quantity ». A few couples from well chosen parents are of more worth than a lot of middling birds.

When young birds are weaned, something of great importance must follow ; adduction.

Often you hear of young pigeons, getting lost even when they had been flying with the same group for several days. Some say, it is because they lost hope, others that the pigeons were not clever enough to find their dove-cot again. The latter are sometimes right; especially if the youngsters have not come from perfect parents.

When you will have read in this book, how to mate your birds, such a thing will not happen to your youngsters again.

Many fanciers let out their pigeons when they are still too young. They go far away from their dove-cot, and are not able to return through loss of strength, etc.

Follow my advice and you will lose fewer pigeons.

First, why are the young birds not able to find their way again, and why are they lost ?

The reason is that being too young, they are not yet capable of finding the way alone.

As is the case with everything that lives, so is it with the pigeon. Its qualities develope with growth, and there is always a stage at which it cannot employ its faculties efficaciously. The principal thing to know is; when can it really employ the qualities which nature has endowed it with, to the best advantage ?

When very young, they cannot find the cardinal points, and so, many get lost.

I have noticed when the pigeons were on the roof, that there was a remarkable change in the formation of the eye, and during that period of development a great many were lost. When the eye begins to develop, the sense of finding the cardinal points changes too.

You will know from the following when these changes take place.

So ever let your pigeons go freely on the roof. Something might frighten them away, and you might not be sure to get them back again.

Build an aviary big enough for your youngsters, and when they have lost their second flight quill, you can let them go out.

But do not pull out this feather yourself ; it must fall off.

Let them out through an opening in the roof. This opening must be placed flat on the roof, so that your pigeons need not open their wings, when going into or out of the aviary.

At this time, that is to say, when they are moulting, the influence of the growth has departed and they can find their way more easily. Their flying power is greater and they know the surroundings perfectly and can recognize the different noises or sounds. So they will be quite accustomed to the spot.

If you see that the youngsters do not lose the first feather, do not pull it out, on the contrary, watch the tail. It is divided into two parts, each containing six feathers.

Beginning in the middle, pull out the second quill on each side.

That helps the moult. Then the first quills of the wings fall also, and the pigeon will be able to find the roof easily when it flies away.

You can be convinced of what I tell you. Try this experiment, you will find it successful.

Now I shall speak of the building, and of the management of a breeding pigeon-house.

THE BREEDING DOVE-COT

Into a breeding dove-cot you must put the reproducers and the breeders.

The racing birds must not be put with them, for the method employed for the former is not tne same as for the latter.

The nest is to be built : 80 cm. high, 60 cm. deep, and 40 cm. wide (91 cm. = 1 yard).

The height is to be divided into 2 parts of 40 cm. each. The upper part is for the mating period. It must have a little railing of galvanized wires of a 1/2 cm. in diameter.

This railing is horizontal and prevents the pigeons in the top nest from fighting. It is hung from the ceiling.

There will be swing doors fixed with the little hinges in order to open inside the nest.

They are to be latticed (trellis) and made with galvanized iron wires of a 1/2 cm. in diameter, placed vertically with a distance of 5 cm. between each wire (5 cm. = 2 inches), so as not to injure the wings of the pigeon.

This nest is chiefly for the cock, when the hen is hatching the eggs in the lower nest.

This nest should be 60 cm. deep.

Always being used for breeding, there should also be a door in the front ; one part of this door should be latticed and the other entirely of wood.

The latticed part is to open on the inside ; the other is only used when cleaning the nests, and it should open on the outside.

This half door must be kept shut during the breeding season, not to have too bright a light in the eyes of the youngsters.

Over the nest, a latticed fence should be placed to prevent the pigeons from remaining there.

You must be careful not to oblige the parents to take their youngsters out of the nest too soon, for when they begin not to take care for their youngsters any more, they suffer very much and sometimes you have to wait till the end of the moult to have them in good form again.

The nest should be as near as possible to the floor to avoid accidents, when the young pigeons begin to go out.

They will be accustomed to the place where they are fed with the old ones. And soon they will be able to eat and drink by themselves.

Some lime dust should be shaken in the nest at least every eight days, to kill the vermin and prevent the pigeons from being annoyed with lice during the time that it is necessary for them to be as quiet as possible.

The nests must be very large, containing a layer of sand and earth about 5 cm. deep.

This will make the nest soft, so that when the hen lays her eggs, they will not be broken. This layer of sand will also help to keep the sternum in its right position. Sometimes, when the youngsters had a bit of something hard under them, it caused their bones to grow crooked, because when the pigeon is young, the bones are very soft.

This sand or dry earth may be kept in the nest till the young pigeons are about 8 or 10 days old. Then it should be changed at least once every three days.

Of course their droppings smell very bad and is naturally, neither beneficial for the young pigeons, nor for the old ones.

The breeding nest must be throughly cleaned and disinfected.

The drinking vessel. — Very often you can only buy earthenware drinking vessels. But a better system is to take an ordinary glass bottle, turn it up side down in a cup, and as it is needed, the water will descend slowly by itself. Put it into a box open on one side with a lid slightly sloping in order to prevent the pigeons from resting on it and making the water dirty.

It will be placed in a light place and before the entrance of the nest if possible ; so, that when the youngster comes out they will get accustomed to go there and drink.

In this way your drinking vessel will always be clean for you must change the water very often. Do not forget to see if the water is always clean and that there is enough.

Add two lumps of sugar or a spoonful of honey (a tea spoonful) to a litre of water (a pint = 0,567 litre).

The sugar will help the formation of the bones and the honey is very good for the organs of respiration.

Begin mating the pigeons from the first of March. Before that time take care to disinfect the dove-cot thoroughly.

I should advise you to give twice a week, from the first of January to the first of March, an infusion of sarsaparilla (smilax medica) and Peruvian bark (quinquina succiribra) as a drink.

A soup spoonful of each to a litre of water. Boil for 5 minutes. Then add a spoonful of pure honey or five or six lumps of sugar. Wat till it is cold and then give it to the pigeons.

This infusion will act as a purative and will have a great effect on the shedding of the down.

The other days it would be a good thing, to give them some fresh water with a few drops of tincture of iodine (tinctura iodii) in it.

FOOD

Very often the food of the pigeon is not the proper sort, and so the health of the birds suffers and leaves them unfit for racing.

What you must never do is to change the composition of the food suddenly. Change it by degrees. For instance you put less and less each day of the grain you do not want to give any more.

From the first of January to the 15th of February, you should give the following food.

30 % wheat, 30 % of oats (small grains), 30 % of rye, 10 % of inseed.

The quantity of food is about 25 grammes a day for one pigeon.

It is to be given in two parts.

Add some cut greenstuff, salad, endives, green cabbage, chicory, water-cress and some grass.

In Winter it will be easier to feed them with potatoes cut into small pieces and beetroot. You may rasp them before giving to eat.

From the 15th of February to the first of March, which is the mating season put big grain into the food and continue that till the first of September.

From the first of March to the first of September, the food will be composed of 25 % of horse-beans, 20 % of peas, 20 % of wheat, 10 % of maize, 5 % of rice, 5 % of linseed, 5 % of new breed, 5 % candied sugar and 5 % of rape-seed.

During that season you may feed your pigeons even three times a day ; never give too much grain, but be sure to give enough.

Never let any food remain in the bins, for it is better to let them feel a little hungry ; it makes them tamer.

They will know the hour at which they are fed and they will come back quickly.

From the first of September to the 13 of December, give plenty the following food.

35 % of horse-peans, 10 % of peas, 10 % of maize, 10 % of wheat. 10 % of rape-seed, 5 % of hemp, 5 % of buckwheat, 5 % of linseed and from time to time some new bread with cod liver oil.

As a drink, give an infusion prepared as follows :

During 10 minutes, boil some dry plantain (Hantago medea), which you find in the ditches, (leaves, branches and seeds) in a litre of water, with two or three lumps of sugar let it cool and give it to the pigeons.

HYGIENE OF THE DOVE-COT

To have a good breeding dove-cot, it must be made of material that will keep out dampness. You can use wood or bituminated pasteboard in preference to eternit which is too cold in Winter.

Pigeons must have plenty of air and light. A pigeon breathes 1 M^3 of air in twelve hours (a M^3 = a cubic metre, is more than a cubic yard for a cubic yard is 91 cm.), and if you have twenty birds, you must give them 20 M^3. But that is the minimum, because the litter etc. makes the air bad.

To have good air, you must first of all, build dove-cots large enough to have good ventilation. The ventilating hole should be near the roof, free from wisps, straw, ceiling or anything else which prevents the air from coming in easily.

A hole should be made over the entrance to ventilate the room.

There must be no other hole, even the key-hole should be plugged. Only the entrance hole and the ventilating hole are to be free to let in air.

The lighting : If I insist upon the fact that the air must be good and plenty of it in the dove-cots, I insist also on good lighting.

A good thing is, to put glass tiles or use a garret-window. The rays of the sun can then enter freely and will heat the interior, etc.

Site : Care must be taken to have the entrance to the dove-cot facing South ; your success will be greater and surer, if you follow carefully what I tell you. The entrance should also be facing the South-East.

The sun's rays will enter early in the morning, and heat and dry the dove-cot.

In this way, many diseases can be avoided, such as a cold in the head, rheumatism and other diseases, provoked by dampness and each of. warmth.

Notice : There should always be a little bin with grit ; little pieces of bricks etc. in the pigeon-house ; but never anything coming from the ceiling, for they may contain hairs, and this would be very bad for the pigeons.

Cleaning : A good fancier will never enter the dove-cot without having the tools necessary for cleaning it. A few minutes every day will keep the dove-cot quite clean.

In doing so, you destroy all that is injurious to the pigeons, such as microbes which multiply rapidly.

Do your best to keep those away and you will save much trouble.

Spécial care to be taken with the reproducers and the youngster of the same year.

Producers should be fed as I have already explained ; but when they are only hatching, give them crumbs of bread till the youngsters are 9 of 10 years old.

The bread which is often called « black bread » or « horse bread » is preferable to that of the baker. It is often made with rye flour.

From that time, they must have plenty of food and be fed early in the morning. In such a way, the youngsters will be well fed, and will allow the parents go out for their daily flight. They will sleep in the sunshine all the morning.

The nests of the youngsters must be free from little insects.

When you see that a youngster is not growing and if it cries even when it has plenty of food, just kill it.

You should also kill it, if it suffers from any defects.

A sound young pigeon which has grown well, must be weaned after 23 days. In this way, the parents are set free, and then it is old enough to feed itself. Put it into another place, and look to see if it gets beaten by the other pigeons. If thirsty, take it to the drinking vessel and push its head into it. After two or three times, it will drink by itself.

As soon as it is weaned, the youngsters must get the following food, and as much as they want of it.

20 % of horse-beans 20 % of peas, 20 % of wheat, 20 % of maize, 10 % of new bread, 5 % of buckweat, 5 % of candied sugar.

From time to time, give them small grain, with some rice, hemp and linseed.

During the first days, it happens very often that the youngsters are lazy. If you discover this ; put a few horse-beans, to steep in water with some salt, leave for about 24 hours, and then give them ten every morning and evening for a few days.

Immediately afterwards push the beak of the bird into the drinking vessel to accustom it to drink.

After two or three days, the little pigeon will drink by itself.

Every evening, examine each youngster. The crop must be full. The fancier who has a special dove-cot for the youngsters of the same year should put two or three old pigeons with them when they are hungry.

Seeing those eat, the young pigeons will eat too.

This dove-cot must be supplied with the same stuff as the others : such as, grit, greens, etc.

A sound youngster will lose its first quill in the wing about 10 or 15 days after being weaned. It must be healthy and lively.

If you see a pigeon make as it were a ball of itself, or if ir doesn't look well, do not wait too long before giving it some little balls composed of new bread, of broken hemp-seed, little brick dust and unsalted butter. (The same pro-portion of each.)

Give it four or five of these balls in the morning and evening.

Then let it drink. The next morning, the crop should ʰ empty .

Continue this till your pigeon is lively again, and able to eat.

The youngsters a year old cannot always find the drinking vessel. You must take care of them, for it happens sometimes that they do not drink and become very weak and may die in a few days.

The pigeons in the dove-cot should be free all the day. They can fly about as they like.

Many fanciers do not keep their cocks and their hens together when they are breeding. To do this, they often close up the hens, and so injure the health.

I shall have pleasure in telling you of a good plan, by which you need not have any eggs laid before the month of March and still give plenty of food to the pigeons.

By this method, you may be sure, that the first egg laid when mating will be the first from that couple during that year.

It is always umpleasant to have hens that lay during the resting time. All their strength goes away, and that without any profit.

These instructions will be given in the fourth lesson, treating of various indications and secrets.

CONTENTS

THE
FOUR SEASONS

REAL COURSE ABOUT PIGEONS

Published by

M. Joseph HEUSKIN

EXPERT IN PIGEON-BREEDING

Léopold Street, 88, FLÉMALLE-GRANDE - Belgium

Translated from French by Aug. LEMMENS

Course in four lessons, treating of the instructions about pigeon-house managing.

SECOND LESSON

Pigeon-lofts — How to feed the pigeons — Home - Coming instinct — Method of managing lofts for widowhood and partial widowhood — Exercises concerning these points.

Course, Number 4522

Pigeon lofts

It is easy to understand that as time advances science too is making more headway, and its application in the practice of pigeon breeding becomes more complicated, especially when a fancier desires to keep his place as a reputed pigeon- breeder.

The stakes become more and more important the races are being organized much better and are repeated more often.

All this causes much anxiety to the pigeon-breeder and makes him study the best methods to be employed.

First of all one must know how to choose good pigeons for racing, and then the special care they need to keep them fit, and able to compete with success in the different competitions.

It must be borne in mind that not only good breed is essential, but also a thorough knowledge of how young ones are to be reared, and the way that the pigeons in which all our hopes are centred, must be treated, and managed.

Above all you have to breed them under hygienic conditions, and in places most favourably suited to the promotion of success — which is naturally your great aim — otherwise you will be sure to meet with disappointments.

To conclude ; a pigeon loft to be properly constructed, must be built on the following lines.

First let me state the hygienic conditions which are of primary importance if you want to succeed.

HYGIENIC CONDITIONS

I. Site of Pigeon loft.
II. Material to be used in its construction.
III. Situation and entrance.
IV. Ventilation.
V. Lighting.
VI. Cleanliness.
VII. Disinfection.

I. SITE OF LOFT.

A very important point to be considered is an easy drop (fall) when the pigeons come home from the race access must be thoroughly studied so that the drop can be effected without the slightest difficulty.

I consider that an attic, if one can dispose of one, answers the hygienic conditions very well.

The breeder who has an isolated spot at a little distance from the bulding, can construct an aviary, or little house. In my opimon that is the ideal loft which will suit all purposes best.

A pigeon-loft situated in a corner without air and light, near a stable or lavatory, over a forge, or any other such place where the sun's rays cannot penetrate is most dangercus and injurious to the pigeon's health.

II. MATERIAL

The material best suited for the interior of the aviary, which you can fix as, and where you like — the outside should be in éternite the to inside in « veneer », applied to all the rafters of from 7-8 cm2, allowing a space exactly the thickness of the rafters which you fill up with quicklime.

In this way you need not be afraid of any dampness. Same temperature Winter and Summer in the loft and you will not have any insects in the crevices of the partition.

For the interior of the house thin boards, « bitumed cardboard » (tar paper) and triplex sheets, commonly called « Plaques Belgica ». The latter are composed of chalk, lime and plaster, they are porous, airy and light, and exist in several sizes, they can be nailed and arranged as one wants without any difficulty.

This sort of material is very advantageous and hever gets damp.

Concrete, eternite, and wire netting are to be strongly disapproved of for the interior of pigeon-lofts.

III. SITUATION

As explained above, garrets and outhouses are the best places for pigeon lofts ; but in order to be a real success you must observe the following rules : ―

The entrance must face the south; if that is not possible it can be south-east ― but in this ca : she pigeons will take a much longer time to come into form (to be trained) than their companions whose entrance faces southwards. The same may be said of all lofts where the opening faces any other point.

Proprietors of such lofts can have prize-birds but not so readily as their contemporaries whose entrance is properly situated.

This point is preferred for several reasons :

1) The pigeon-loft facing south gets the first rays of the sun. Thus it is heated and kept dry. The sun being the best disinfectant, of the material used in the construction of the loft, and also for the pigeons ;

2) It is sheltered from wind and rain, especially from the harsh north wind which is so injurious ;

3) Then a very important point, but one which many fancier ignore is that the birds bred in lofts facing south are much better than those at any other side.

Constant observation during several years experience in pigeon breeding has given proof of all I have said to pigeon fanciers in the first lesson about pigeon breeding, therefore I know that you will obtain much better birds and you will prevent many diseases such as coryza, arthritisme wing disease, etc.

IV. VENTILATION

When you have to make a pigeon loft you must never neglect proper ventilation.

Within the past few years a league has been formed giving full particulars concerning the construction of places intended for stables, pig-sties, and the housing of other animals on the farm. This league especially insists upon proper ventilation.

Do not forget, dear readers, that pure air is as essential to the pigeon — or more so -— than to any other animal on account of contagious diseases which can be contracted in the baskets, etc., then once more in their own appartments, they must have everything they require, and fresh air is of primary importance, as the health of a whole colony can depend solely on the air they breathe.

A well planned pigeon-loft has only one air-passage, that is through the entrance, and the air must go out by a hole just directly above the entrance. There must only be those two openings in the whole pigeon loft. All the other parts must be air tight — no cracks anywhere, even the keyhole must be plugged.

A tile roof is the most preferable for a pigeon loft — let it be either a garret ar a « swiss cottage » flat tiles are best. This allows of the most excellent air-passage.

A pigeon requires 1 cubic meter of air per day. Generally, dovecots are much too small for the number of doves kept there. Thanks to the ventilation hole and the flat tiles, they can have sufficient air.

When the roof of the pigeon-house is plastered over or lined with some other material it obstructs the free passage of air and thus causes trouble in the colony.

V. LIGHTING

So that the pigeon-loft be constructed under hygienic conditions it must have plenty of sun and light.

The direct rays of the sun must be able to penetrate to the interior.

If necessary you can put glass tiles or sky-lights on the roof, so that the very best agents to health — light and sun — can penetrate from the top. Compare a gloomy appartment with a well lighted one — the latter is much more pleasant and has the benefit of the ultra-violet rays.

The sun is a great antiseptic. A badly lighted pigeon-house is what makes it often so difficult to rear the occupants, and leaves them much more subject to disease than pigeons bred in well-lighted and sunny houses.

VI. CLEANING

A pigeon-house and all its accessories should be cleaned every day, and be thoroughly disinfected at least twice a year.

A fancier who really loves his colony (group) is always anxious about their interests — he will not give grain without having previously cleaned the place, — he will do so before every meal, first making sure to remove the drinking water until he has finished cleaning — this is to prevent dust getting into it.

Do not imitate many of our negligent fanciers who wait until there is a layer of droppings two or three inches thick on the floor before they think of scraping it off. These droppings stain the floor and leave a most unpleasant smell, it is very injurious to the pigeon's sight, and the noxious odour which fills the air is very destructive to the health of the birds.

Do no forget that when a bird catches some infection, the germs are always in the droppings, therefore if care is not taken the infection can spread through the whole group (colony) — and through drinking the same water also. A bird that has contracted some infection during the race may not be suffering on its return, but can spread the contagion to its neighbours. Pigeons kept in dirty houses breathe bad air and can therefore contract diseases which may become contagious or chronic.

DESINFECTING

The periods for complete desinfection generally is (for various reasons) before pairing the pigeons, and after the races.

To make a thoroughly good job of it, it is necessary first to remove all the birds to other quarters or to put them into the racing baskets. Then pull down all the inside, leaving only the four walls, clean all the dirt up, take away the gutters; replaster; and whitewash thoroughly; also clean all utensils belonging to the pigeon-house.

The desinfectant which I recommend for this purpose is a mixture of thick whitewash, to which you add 1 kilo. of table salt 250 grammes « cresyline »; 250 grms. of soft soap; ½ litre of petrol; ½ litre turpentine; 0.5 grms. of « Formol » — making about 10 litres when all is mixed together — i. e. an ordinary bucket-full.

You should give two applications of this — going into all the holes and corners, not forgetting a single one.

This work being carefully gone through, leave to dry for about 48 hours, then air well, you may even cause a draught before putting the birds back again.

The fancier who will carry out these instructions can be sure of having healthy, pigeons, and will prevent trouble in his pigeon-loft.

If a contageous disease breaks out in the pigeon-house, it ought to be disinfected without delay according to the directions given above, and to be sure of destroying all the microbes, amongst others, those hidden in little cracks and crevices it is excellent to fumigate the premises.

For this, one must close the doors and windows tightly and stop any holes or cracks — in order to render the pigeon-house air tight. For the latter, purpose one can use papers, old rags or bags.

Then one places a vessel containing sulphur (powdered) about 50 grammes per cubic centimeter of the pigeon-house; over this, one sprinkles a little alcohol.

One must be careful to set the pan on bricks or stones, this done a lighted match is put to it and the person must rush out quickly, closing the door tightly, by stopping all slits a openings with old rags.

The sulphureous gas spreads through the whole place, penetrating everywhere and killing any insects or microbes which may have remained after the 1st disinfection.

After a certain number of hours all the windows and doors should be opened to air the place thoroughly.

Remark. — It is comprehensible that all the birds must be first taken away from the pigeon-house.

GENERAL CONDITIONS NECESSARY
FOR A GOOD PIGEON HOUSE

A well planned pigeon-house, must have the best qualities with regard to structure, ease, and comfort.

Access to the interior must be quite easy — for persons as well as for the birds.

To obtain the best sporting results it should be arranged as follows :

1) Access to the pigeon-house ;
2) Observatory ;
3) Roof ;
4) Precautions against dangerous surroundings.

1) ACCESS TO PIGEON-HOUSE

The easiest possible access must be to the pigeon-house. The stairs leading to it must be firmly fixed. The steps should be quite horizontal — a strong life-preserver schould be adapted in order to avoid accidents when in a hurry, or in Winter during snowy weather. Needless to say these conditions apply to highly recommended lofts.

The floor has to be very strong — no slits or chinks ought to be seen.

The entrance door must be at least 75 cm. wide and sufficiently high to allow one to pass in easily.

At a certain height, and preferably in the middle should be fixed a pane of blue glass, through which one can see everything that goes on inside without being seen by the pigeons. A pigeon-house to which access is not easy is the

cause of many accidents, and toss of money at the time
of the races.

2) OBSERVATORY

Each fancier must be able to observe everything that
goes on in the vicinity of his pigeon-loft.

An obsevatory is a most useful equipement for any pigeon breeder. He who can dispose of the necessary space
— can construct an observatory big enough for himself and
several other persons if necessary to install themselves
with ease.

To be a well installed obscrvatory, there must be place
enough for at least four persons. It can therefore be com-
pared to a little house overhanging the whole pigeon-loft.
It should be made of wood or iron and have four windows
which can be opened and closed at will.

There must be a blue pane in each and also one in the
roof. It is through these panes that you have the opportunity of observing everything, and from each point, without
being seen.

Then you will have the satisfaction and pleasure of
perceiving at a long distance, the arrival of you favourites
from the race.

If, amongst your group, you have a good racing bird,
but one that is capricious, and on returning from the race,
flies over the loft, or remains on the roof instead of entering immediately, it can be perceived in time.

Being inside your observatory you can immediately go
out through the window which leads to the entrance of the
pigeon-house — this exercise having been practised so as
to be accomplished without difficulty — there will be no
cause of delay in bringing the racer into the loft.

Then in case of bad weather when at your post you are
sheltered and have nothing to worry about with regard to
your health.

During their daily flights, what pleasure you will have looking at your pets during their sport, thus you can take note of many important facts which will high.y compensate you during the competitions.

At the breaking in of the young birds or of a new subject, you can easily see where it goes, and act accordingly.

Every day when letting them take flight you can go to your observatory and see if anything frightens or troubles the pigeons, for instance a slater on the neighbour's roof ; a cat; or anything else which might be of danger.

Useless to insist on the facility with which access to your observatory must accomplished.

There ought not to be the least difficulty, and if a stairs is required, there must be a strong life-preserver in case of falling down — and the stairs must be horizontal.

It can also happen that a pigeon enters unnoticed therefore place your observatry so that — when seated you can easily see into the pigeon house — especially if there be an opening in the wall.

It may be that one has not sufficient space to construct an observatory such as I have described, then one can make a similar one only smaller. — which must certainly have panes of blue glass on every side as well as being constructed so that there is no difficulty in going in or coming out.

3) THE ROOF

There are many kinds of roofs, but most of them are quite unsuitable for pigeon-lofts.

The one which I favour — from every point of view for pigeon breeding. is a tiled roof (tiles without hollows) as there is a free air passage while at the same time the roof is tightly closed. And access to the roof for repairs or any other work is quite easy.

Roofs made of slates ; eternite ; zinc ; tar-paper concrete, etc., cause difficulties, and are injurious the health of the inmates.

A slate roof lined inside with boards prevents the sun and air from penetrating ; the same is to be said of the roof of eternite.

A zinc roof is very unhealthy, in Summer it is too hot and it suffocates the pigeons as it does not allow of the air being renewed. In Winter it causes dampness, which is equally unhealthy.

The roof in tar-paper must also have a lining of thin boards and then in order to preserve it. A coating of tar must be applied at least once a year.

During this operation and for some days afterwards your pigeons should stay in the loft, this is very important for a bird which flies on to a tarred roof, newly coated may spoil its wings ; and what is quite certain its toe-nails will catch in it and may even get pulled off.

The concrete roof is out of the question on account of its being so damp, because the pigeons receive no heat whatever from the sun during the whole year.

So that the roof of the pigeon-loft be well constructed it is necessary that it be provided with glass tiles and skylights. The latter ought to be placed as near as possible to the carnice and the entrance : however you must take care not to place them directly over the pigeon-holes.

These sky-lights placed in proximity with the carnice and the entrance, leave a little opening of about 5 centimeters, night and day, and in all weathers so as to draw away any bad smell from the pigeon house. In one of these skylights there must be a cage with a wicket or an opening large enough to be to reached from one end to the other with the hand : this enables you to instal the young birds for some hours every day when you are breaking them in and when you must break in a new bird to your group it will be very useful.

This cage will undoubtedly be of great service when you have a good racer, that does not enter, but stays on the top of the roof for some time after coming home.

In the fourth lesson I shall give you full instructions as to how to make them go in quickly.

The carnice is genarally supplied with zinc sheets, which collect the rain water from the roof. Zinc being a deadly poison for the pigeons, it is more prudent after a drownpour of rain to see that no water lodges there. Some people keep this water for their pigeons to bathe in, such a proceeding is to be strictly avoided.

If the entrance to the pigeon-loft is through a window or glass frame in the wall, there must be a zinc or galvanized iron tube 10 or 15 cm. in diameter laid through serving as an outlet, etc., for the bad smell from the pigeon-house.

When the entrance is through the roof, a hole must be bored in the wall directly underneath it and as nearly as possible on a level with the ceiling the outside must be provided with a lattice in which there is a swing-door (dimensions 30 x 25), hung so as to be able to open on the inside. The pigeon-house can be kept constantly dry with this little window, and at the same time it will enable the renewal of air when necessary.

If the roof is one upon which the pigeons can easily rest, or if the slope is not sufficient and the entrance is through an opening in the wall, arrange the cornices in the same way as the sides, with a partition, (treillis) about one meter high, this will make the pigeons fall directly on to the landing-board. In this manner they will acquire that habit, and on their return from the race will win time and money for you.

If you are obliged to place your entrance in the roof, what will be most satisfactory is a glass frame or « chassis flamand », the shutters or door must work on rails or ball-bearings, or on little rubber wheels.

When the roof has to be repaired, and the work must be done in mortar, always use cement.

The first year, ordinary mortar is bad for the health of the pigeons and causes slow poisoning.

When the work is complete, keep your pigeons still closed in for two or three days after that you can let them out but prevent them from going on the roof for some time yet.

4) DANGEROUS SURROUNDINGS

Things which are to be greatly dreaded, and which cause any amount of trouble, when approximatively about the same height as the entrance to the pigeon-house are : kitchen chinmeys; furnaces; factories, etc., etc.

The gas which escapes from them is poisonous and therefore most injurious to the health of the birds. When the wind b'ows in the direction of the loft, gas and smoke enter and make the whole group sick.

I do not know any means of preventing this catastrophy except to change the site of the pigeon-house to another spot, or when possible to raise the chimney higher than the roof of the pigeon loft.

As a precautionary measure all chimneys to which the pigeons have access — the young ones in particular — ought to be covered with a wire netting or lattice to prevent the birds from falling in when they begin to fly.

If there are telephone-wires or any other wires which might cause accidents near the entrance or about the loft, the simplest way to prevent such accidents is for you to place a second wire close to the others on which you must thread quite a number of corks, or other very visible stuff so as to mark out these dangers.

In striking against the wires they can become unable to re-enter their loft, through bruises, fractures, or perhaps broken feathers.

If also around your pigeon-house, there exist any buildings of unequal height, which could serve as a ladder for cats, be careful to fix up a trellis high enough to prevent them from getting into the pigeon house.

QUTERIOR OF THE PIGEON LOFT

In the preceding chapter, I spoke to you about the outside of the pigeon-loft; how it must be arranged, what precautions to take, and the material to employ — also the material which must not be used.

In this chapter I am going to draw your attention to the interior, and all its chief points.

THE DIMENSIONS ACCORDING TO YOUR SYSTEM OF THE SPORT

Let me point out to you, that, if you have place enough the dimensions which I describe, must be carried out. As it is not always possible to have what one desires one must do as best one can, but to have the greatest chance of success, it is first necessary to study the facts indicated.

What I wish above all to impress upon you is that a pigeon house must never be too large, for the reason that the birds become wild : on the contrary birds and master should be friends.

Is there anything more disagreable than to have untame pigeons, since familiarity with the birds is necessary in pigeon breeding ?

The dimensions which I consider most suitable for both fancier ands birds are :

1) length : 2,50 metres to 3 metres ;

2) width : 1 ¾ m., between the partition opposite the pigeon-holes — and the pigeon holes ;

3) height : 1 m. 80 cm. to 1 m. 90 cm. maximum.

If the group is small, reduce the lenght to that necessary.

ENTRANCE

The entrance to the pigeon house is the part of the pigeon loft which needs most attention it must be combined in such a way as to make the birds fond of it; to be able to enter without difficulty; it must be supplied with all accessories according to the play you wish to take part in.

The entrances to pigeon-houses differ very much one from another.

In a pigeon loft where widowhood is practised the entrance must be much bigger than that of a pigeon loft where only breeding is carried on.

In an installation where speed flight is practised there are always as many entrances as there are pairs of pigeons.

In the following chapters I shall give you particulars necessary for the construction of entrances according to the play practised, and at the same time other important information.

THE PIGEON-HOLES

The pigeon-holes must be made of polished wood, no cracks are to be seen, and made so that they can be taken to pieces; the sizes vary according to the way in which you train or manage the birds.

The position, means of shutting, and the partition differ according to the object for which the pigeon loft is intended.

In the pigeon-house where racing is practised the pigeon-holes are raised from the floor and extended to an open-work partition underneath, in order to prevent the pigeons from going into this hiding place; for when you must stoop down, or climb up on something to catch the birds after the race it would cause much delay.

Under the pigeon-holes and behind this open-work partition, — arranged so that the birds cannot get there, — you must scatter a layer of lime dust about 5 or 6 centimetres deep, this keeps the pigeon loft dry and kills all the microbes.

In pigeon-houses for breeding the holes should be placed on the floor.

EATING TROUGHS

The grain should always be put into troughs. During the racing season the pigeons get all their food in the pigeon-holes. It should be given in little earthenware vessels (troughs).

In this way you can give to each bird the proper sort of grain, in order to prepare some for a certain race, and maintain others for longer periods.

When giving instructions about feeding. I shall have the pleasure of pointing out to you, of what the rations are to be composed, how a pigeon must be fed in each season, according to the sort of race for which it is designated.

During the resting season the food can be given in one trough, which be can made out of a flat board surrounded by four little thin boards about 2 centimetres high, these keep the grain from falling on the floor.

In order that each bird can take its place at the trough the length ought to be 10 centimetres for each pair of pigeons, eating from the trough.

Do not employ a trough with a lid nor one separated by laths.

If I recommend the trough with rails around it that is because it is the most advantageous sort.

First because it can be cleaned more easily. and the birds will become accustomed to each other and will not be afraid when put into baskets together.

Generally a bird in the habit of eating alone will get afraid if pecked by another and go away and not eat any more, to its detriment especially for long distance races.

DRINKING VESSELS

Many kinds of drinking vessels exist but they all have their defects.

Some are most injurious to the health of the pigeon, others are difficult to clean, and are therefore unhealthy.

The one which I consider, as being the most advantageous, is an ordinary white glass bottle of about 1 litre.

You must turn it upside down in a large cup and attach it to the partition with a pièce of wire, cover the top so that no scum or dirt can enter.

The drinking vessel should be placed opposite the entrance, for several reasons.

First it will receive the air coming in, and then being placed in the light you can easily see the contents; it will also induce the birds to enter more quickly after a race, for they are always thirsty.

This system will oblige you to change the water every day and will help you to ascertain at first sight when the vessel is empty.

A litre of water is sufficient for one day, when the temperature is normal and the group is not too large.

During hot weather, and when the birds have their young ones, it is necessary to look often at the versche, because they drink a great deal. Zinc drinking vessels are very dangerous, the metal coming in contact with the air and water dissolves and becomes a deadly poison : others are difficult to clean.

Sometimes you see drinking-vessels which coutain 5 or 6 litres, and which are only washed each time they become empty, and these are employed only for a small number of birds. Hour can it be otherwise than that the pigeons absorb dirt and all sorts of miasma which is blown about when the birds flap their wings. It sometimes happens even that on entering the pigeon loft, one gets a very bad odour from the drinking vessel, needless to say the water cannot be good.

You must see that vessels made of iron, brass, or earthenware be cleaned every day, a little pièce of cloth will do te clean the inside.

BATHS

Baths are as necessary for pigeons as for people. If it is necessary to eat and drink, it is just as necessary to take a bath.

You must therefore make it bathe sufficiently, first for its health, and then to prepare it for racing.

There are baths made of earthenware, galvanized iron, zinc, etc., which should not be used especially those made of zinc, because the metal slowly decomposes through the effect of air and water and thus becomes oxidised. This oxidation transforms the water into (a deadly) poison.

The same thing may be said of those in galvanized iron, although the danger is not so great.

Earthenware baths, although not injurious to the health have their disadvantages, for the reason that some of the birds will go directly to the water, while others would like to do so but are afraid ; then again there are some pigeons that do not like to go at all, while their companions stand on the bath with one wing spread out, quite satisfied with the splashes they receive from those in the water.

The ideal bath is an ordinary pail, and if you follow my instructions, all the birds will be sure to take their bath.

The water for the pigeon's bath must be lukewarm. Heat two or three litres of water in a kettle, pour it into the pail, then fill the latter with cold water to about five centimetres from the top, taking care to have the whole at about the temperature of the pigeon's body.

Then you stand the pail of lukewarm water in the middle of the pigeon-house, but first put an old cloth or bag underneath, in order not to dirty or wet the floor — which is very bad for the birds.

Then take the pigeon in your two hands, being careful not to pull out any feathers ; plunge it into the pail of water, all except the head, and hold it there for about 45 seconds.

In order to let the water penetrate to the skin you must open out the feathers especially the underneath ones, all along the tuft on the breastbone, under the wings and on along to the rump.

It is in these parts that you find the thick down closely matted together which you must carefu'ly work through. It is chiefly those feathers which hurt the pigeons and prevent them from flying very fast. Let me draw your attention to the fact that when bathing your pigeons, you must keep the door closed until they are quite dry.

After the bath, when they are thoroughly dry let the whole colony take a good flight, to warm themselves.

In this way they get rid of all the bad feathers down and dandruff, etc., etc.

During the racing season the bath must be given on Monday or the day after the race.

The pigeon which is to take part in a competition must not get a bath the day of its despatch.

You must add a good handful of tab'e salt to the water, this is very beneficial to the pigeons, it kil s vermin, prevents perspiring, and is a preservation against congestion.

A bird which is entered for a long distance race must have a bath the day before it is despatched.

During the moulting season, when the pigeons are resting, it is necessary to give them two baths a week.

Before you break in a new bird to your group, it is more prudent to give it a bath, for if it had been badly cared for by its old master, it might cause some trouble in your colony.

If there is a young bird amongst the group which cannot shed its first wings, give baths ; that will help ; the same can be done to a bird that has a stoppage in the moult, without, of course having contracted any disease.

Lukewarm baths is a cure against weak wings and lameness. Baths given to a pigeon worn out from hard work, will soon enable it to take flight again.

ROOSTS

During the resting season when the racing season is over, many fanciers pull down the pigeon-holes and put up roosts or perches.

In so doing they make a great mistake, as roosts have many disadvantages.

They cause the pigeons to develope a defective posture, and deform the breastbone. The birds become excited, peck each other in the eyes, fight and thus destroy their wings and cause additional and unnecessary fatigue.

When the races are over, the set of pigeon holes should be taken down, properly cleaned and white-washed (disinfected), then put in a place exposed as much as possible to the cold, but sheltered from the rain ; until needed for the following race, they can remain like that.

Then they must be replaced by wooden cubes (as is explained in the third lesson) for the dovecots during the moult and the winter season.

NESTS

You can buy earthenware nests of different sizes and models. Some of them are perforated at the bottom, others are not.

The former are not to be recommended for the reason that insects can come in through the holes and take refuge in the nest. They suit neither for breeding nor racing birds. If you have nests of this sort, to be economical, stop the holes by plastering them over.

If you go in for breeding, use the large sized nests without holes. Put a good coating of tar on the outside of the bottom of the dish, and let this dry well before using.

Then underneath shake a good layer of lime : the tar is a disinfectant and will prevent insects from going into the nests, while the lime will keep the bottom quite dry and destroy any germs which night develope there.

As long as the nests remain in the loft, they must contain a certain quantity of earth or dry sand, this is to prevent the female from breaking the eggs, either when laying or sitting on them.

This method keeps the pigeon free from vermin where widowhood is practised — during the resting season and protects it from dampness.

GRIT

In the first lesson I gave instructions about the digestive organs, and pointed out that the gizzard contains plebbes and that it must therefore absorb a certain quantity of mineral matter and lime to keep it in goed working order.

These substances can be composed of ground eggshells, dried in the oven, or also the residuum from bones well pounded into powder. For the latter you must have a kilogram of bones which you put on a strong fire, let them burn until they become quite white then take them out, grind them up, add a few pieces of broken bricks, a few cuttle-fish bones, old mortar and a little table salt.

This mixture must be put into a bin, with a sloping lid, which surpasses the brim, so that the droppings from the birds perched above do not fall into it.

Openings must be arranged on the sides at a distance of about 5 centimeters from each other and constructed by means of a (galvanized) wire about 5 or 6 mm. in diameter. Beside the bin another vessel should be put containing table salt.

Here is a very simple and inexpensive way of making a block of salt, which you know to be free from any adulteration

Take 1/2 kilo of salt, moisten it, and then put it to dry on the mantelpiece or near the fire ; you will obtain the nicest and best block which can exist.

There is also another bin which must be always kept in the pigeon-house, except that of the racers during the racing season ; it is a bin containing chopped green-stuff.

OUTSIDE DOOR

As I have already said the entrance must be on the south side or if possible South-east.

The opening must be very wide, but the height can be restricted. A pigeon needs a large space to descend while it is not necessary to be so high, for it drops much better into a wide low entrance, than into a high and narrow one. These conditions must be adhered to whether the entrance be in the roof or in the wall.

When the opening is not sufficient for the pigeon to enter without stopping, there ought to be a landing board of the size required, so that the birds can stop without difficulty. Thus you can avoid their getting lame, pains in their wings, ruptures, etc.

During my several years experience I have seen many accidents of that kind, while the fancier was quite astonished to see his pigeons losetheir value, and wondered what was the cause : it is quite plain — a pigeon returning from a race worn out, must thrust itself through an entrance which is not big enough ; if it stumbles or slips, it can cause a rupture inside ; or if the wind is too strong it can be thrown against the partition and soon you notice a weakness of the wings of legs.

Different Sorts of pigeon lofts

In the preceding chapter I have described the material most suitable for the construction of the best pigeon-houses for carrier pigeons.

The styles or models differ very much from each other, and so does the material. They are constructed according to the system of sport which is practised, and the material is in accordance with the model.

Some like to practise natural and others partial-widowhood, many prefer speed competitions and so on, and even total widowhood, this needs a special dovecot for the female birds.

First of all one must adopt a special system. Apply onesself to it and build the pigeon-houses for that speciallity.

Every fancier ought to have a special pigeon-house for breeding — this is the one which renders the highest service, and ought te be made first, therefore I continue my course in the following order :

1) The breeding loft ;
2) The loft for natural or partial widowhood ;
3) The loft for widows ;
4) The loft for speed competitions ;
5) The loft for females.

I shall also give a description of the swiss cottage, which in reality is the ideal dovecot because it can be planned in such a way as to practise all the different branches of the sport.

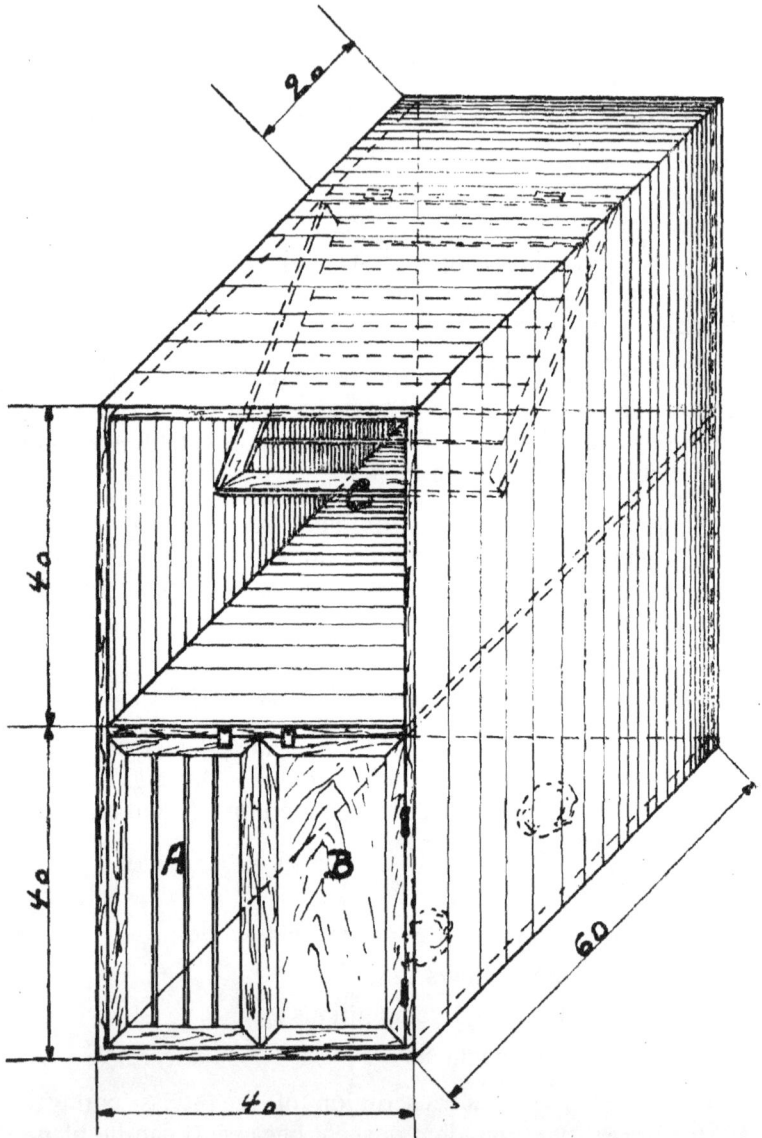

Breeding loft

BREEDING LOFT

The breeding loft must only be occupied by the birds accomplishing this role, and it must have as many pigeon-holes as there are pairs of birds.

The pigeon-holes are to be of the following dimensions height 80 cm. depth 60 cm. width 40 cm. Then again they must be subdivided into two equal parts from bottom to top in order to form two sets of pigeon holes one placed over the other.

You will notice that the female likes to go into a quiet spot while the male occupies the nest, so as to avoid being annoyed by the others. This is described as follows : —

The top part must have a swing-door, hung so that it opens on the inside.

The lower part must also have a swing-door of which one half is of wire netting and the other of wood.

The wire part must open on the inside and the wooden part which is needed when cleaning out the holes and nests must open on the outside.

On the top set of holes you must put another partition of netting, hung from the ceiling by two hinges so that it can be raised up, it is used for pairing the birds.

For this purpose it is preferable to use round galvanized wire about 5 millimetres in diameter, and it should be laid horizontally so as to prevent the birds from pecking eachother.

If the sets of pigeon-holes do not reach the ceiling, you must make a partition with round wire, laid vertically. This is to accustom the pigeons to remain in their holes, and to prevent them from going on the top to fight.

Those are to be turned towards the south or south east, under the doorway if possible and they must have two large nests, such as I have already described.

The drinking vessels, the bin for the greenstuff, the grit etc., should also be placed in the bottom of the pigeon-loft.

The grain can be given in covered troughs, with partitions on the sides.

For further particulars see first lesson on breeding.

———

PIGEON LOFTS FOR NATURAL OR PARTIAL WIDOWHOOD

The pigeon loft for natural or pratial widowhood can be divided into two parts by a solid partition placed beside the entrance and continuing to the bottom of the dovecot.

The old birds should occupy one part, while the other should be reserved for young birds of that year.

The pigeon-holes for the old birds must face the east and those for the young ones must be towards the west

There must be no passage, the entrance is to come within about 25 cm. from the floor.

There must be two openings, one for the old pigeons and one for the young ones.

In order to be able to see everything that goes on inside, it suffices to have a door in wire netting in front at about a meter from the outside part of the entrance. This door, — (with its sides and walls) forms a cage of about one meter in depth.

It has to be made of round galvanized wire of about five millimeters in diameter laid vertically.

There must be a landing board outside the entrance to enable the birds to land without the slightest difficulty or accident.

This door in wire netting situated directly opposite the entrance is to be hung on hinges so that it can be opened and closed at will.

This will allow of clapettes (bobs) on the side occupied by the young birds, and also at the bottom of the wire netting on the opposite side for the old birds.

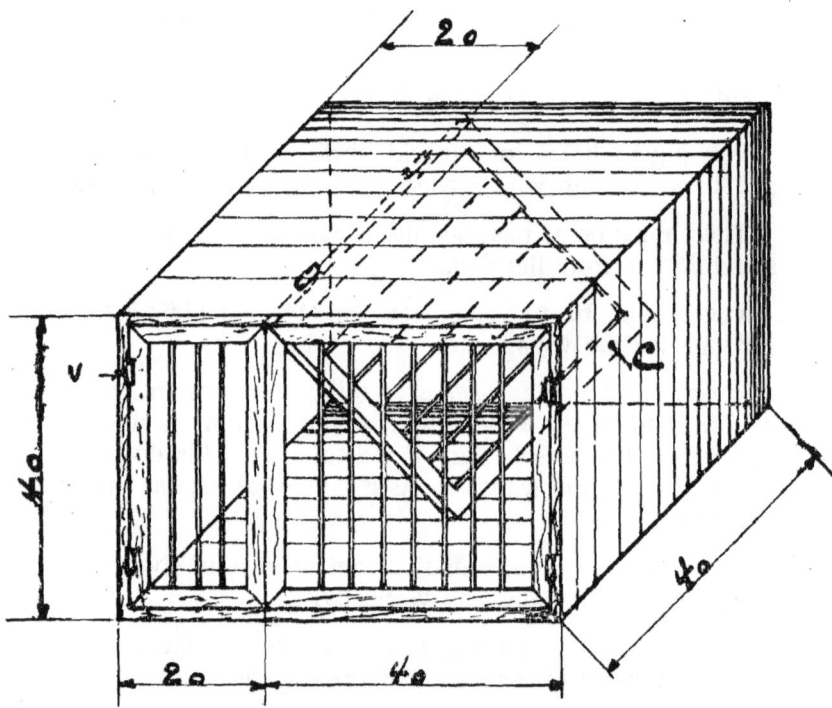

Lofts partial widowhood

A slide with clapettes along the whole width is to be placed at the interior of the entrance at about 50 centimeters away, so as to form a second cage making it easy to take the pigeons belonging to any one of the pigeon-houses.

These bobs must be placed so that the pigeons can go in quite easily, but not be able to come out again.

In the young pigeon's cot there must be a way out either through the skylight or through any other opening in the roof, surrounded by a cage in wire netting, which will help them at the breaking in period, to examine the surroundings.

The sort of shutter which I prefer for a pigeon loft is a curtain or screen, which works from top to bottom ; it should be just at the outside of the loft.

It must be made of opaque material (black or grey linen) so as to cause darkness ?

In the fourth lesson in the instructions about forming the pigeon. I sha.l have the pleasure of explaining the reason of this obscurity.

It is also necessary to place a door for Winter at the entrance, that is to say, a slide with panes of glass ; it must be movable, and hung so that it opens, not from top to bottom but sideways, therefore horizontally.

This horizontal shutter enables you to make the opening as small as you like. In bad, or damp weather you leave an opening for one pigeon to pass in or out at a time.

The panes of glass allow the light into the pigeon-house, the entrance being shut.

As a measure of precaution — especially in case of bad weather it is necessary to place a bell quite near the bobs or the entrance board, to warm you when the pigeon enters.

If it is impossible to place the entrance to your loft, in the wall, then employ the glass-frame, or « flemish entrance « in your roof, on condition to keep to the dimensions

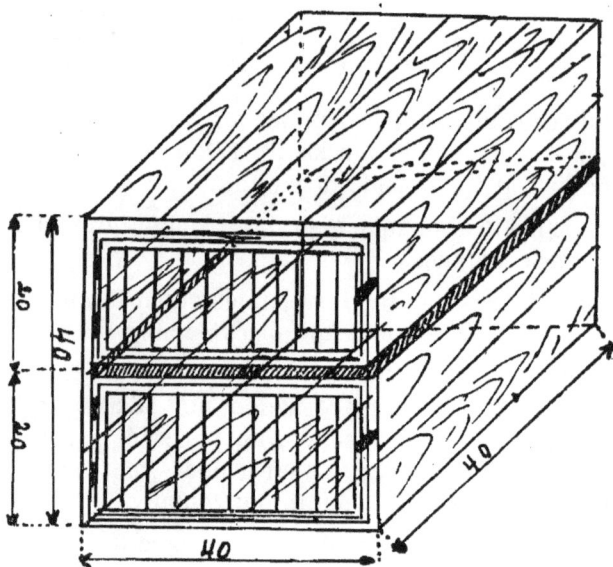

Loft for natural

of the width. It can be set up in the same manner as that situated in the wall — for this it is sufficient to bring the landing-board far enough into the interior of the pigeon-loft.

It is also necessary to have a skylight, or glass tiles on the roof to let plenty of light in.

However, during the racing season, there will be a time when darkness must reign in the pigeon loft, then it will be necessary to put, underneath the skylight or tiles, a slide or sort of shutter to keep out the light but not prevent the air from penetrating.

In the compartment for the old birds the set of pigeon-holes must be placed with back towards the East, and in that of the young birds towards the West.

This pigeon-house containing two different systems of the sport — the pigeon-holes must be of different sizes ; one set is destined for partial widowhood, and the other for natural.

The set of holes for partial widowhood must be 60 cm. wide, 40 cm. deep and 40 cm. high.

The front must consist of folding-doors, one of which is to be 40 cm. by 40 cm. the other 40 cm. by 20 cm. In my instructions about managing a pigeon-house for partial-widowhood, I shall explain to you why there are doors of different dimensions. This place is of great importance, according to the system of the sport you practise.

They should be made of netting — the wires should be laid vertically at a distance of 5 cm. from each other.

Inside there must also be a door fixed in the ceiling so as to form two places in the same pigeon hole, and which can be removed at will, it must be hung on little hinges leaving a space of 20 cm. on one side and 40 cm. on the other when it is overhanging.

The wires which are used for the construction of this door must be laid horizontally 5 cm. apart. These instructions must be rigorously observed ; for this prevents two birds together, from pecking each other or from injuring their wings, fighing.

The holes for the system of « natural » must be of the following dimensions 40 cm. high, 40 cm. deep and 40 cm. wide. They must also be divided into two parts by a partition of boards — from top to bottom, — which can be taken to pieces at will — so as to leave a place of 40 cm. square when taken down, and when set up to form two places 20 cm. high, 40 cm. wide and 40 cm. deep.

The front must consist of two little doors in wire netting or lattice work set up so as to open towards the interior.

These holes are arranged so as to leave two spaces where you can put one pair, but they are two low for them to have any connection with each other.

In the instructions relating to this system of the « play ». I shall give you further details.

In this pigeon-house the pigeon-holes, must be raised 40 cm. higher from the f.oor ; the height must not be greater than your own height so that you will not have any difficulty when you want to catch the birds.

If there still remains a free space on the top it must also be closed in with the same wire and made to form a door, to allow of every facility, to clean and disinfect properly.

Under the pigeon-holes you must place a board to fill up the space and underneath that, you must shake a layer of lime dust about 5 centimeters thick, during all the racing season.

Instructions concerning the steps to take for the management of this pigeon-house will be given at the end of this lesson.

THE PIGEON-LOFT FOR WIDOWHOOD

The pigeon loft for widowhood must only be occupied by the birds (pigeons) designated for this system there must be none but the racing widowers.

It is even advisable to have no other pigeon-house near hand, any pigeons, even the young ones, about the place, do an amount of harm to your subjects.

The entrance must be as big and as wide as possible.

It would be preferable te instal this in the wall, in an end well exposed to view, the dimensions recommended for the system of widowhood are : — 1 m. 20 cm. wide ; by 1 m. high.

Working exactly ou these dimensions, a landing board will not be necessary, only you must not forget, to make the passage which I have already indicated — one meter long — this replaces the landing-board, and enables you to catch the pigeons immediately they enter.

This passage must be raised 25 cm. from the floor naturrally the width of the wall is included in the 2 meters.

If these dimensions be strictly observed, you will have the opportunity of contemplating the ease with which your pigeons will enter ; — a thing to be greatly appreciated.

The outside can be closed by means of a curtain in grcy or black linen ; which functions from bottom to top or vise-versa.

In any case if the passage is situated in the middle of the pigeon-house, or agrainst a partition, the entrance should have the two sides filled in.

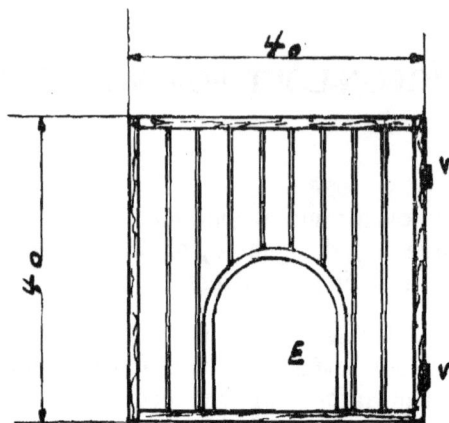

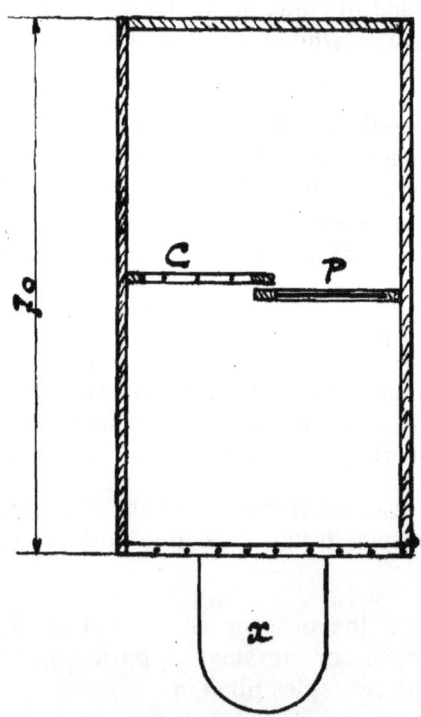

Loft for widowhood

Opposite the hole, the door must be in railing or netting and supp.ied with « bobs », so that the pigeons can go in but are not able to get out again. However do not employ iron rods or little thin boards to hold up the « bobs », make them press against the bottom, the pigeons will soon acquire the habit of « falling » into the pigeon-loft.

When you want to let your birds come out, raise just two or three of those bobs, so as to let one at a time pass hrough, this will avoid their breaking any feathers, in rushing into the passage.

The door must be hung upon hinges and in such a way as to leave the passage free towards the outlet.

The pigeon-holes are to be placed on the right or on the left of the entrance — for various reasons.

Firstly, you will avoid too sudden a change of temperature when the bad weather comes, and secondly you will have great facility in watching how they behave themselves as to coming into form while they are resting.

At the time of the return from the races, the door must be quite open so that each bird as it comes back will have no difficulty in going into its proper hole. When you must go away or when the weather is uncertain, close the door, and set the electric bell to function so as to warm you when the pigeon arrives.

In Winter when the weather is very bad it is equally necessary to put up something to stop the entrance, a window or panel in celluloid, of the same dimensions as the front door, and adapted to it, taking care at the same time to leave a little opening underneath, about the width of two « bobs » ; enough to let one pigeon pass at a time. This panel or window so constructed, will let plenty, of light into the loft while for the most part it closes up the entrance.

In the loft for widowhood, it is advisable not to have the same means of closing the entrance in winter, as that which is used for Natural, as it has no landing board outside. This sort of shutter being placed directly against the wall and only leaving a hole sufficient for one pigeon to pass, the birds can come up against it and hurt themselves.

If you cannot arrange the entrance in the wall make a hole in the roof ; the frame which I have already described, let it project sufficiently outside so as to be able to be constructed according to dimensions indicated for the width.

The roof must be just the same as for the other pigeon-houses, and also the position with regard to the cardinal points (see preceding instructions).

The set of holes must be placed over each other and be of the follwing dimensions 70 cm. deep,. 40 cm. high, and 40 cm. wide (see figure).

At first sight it will seem to you that the depth is exaggerated, but from the following it will be easy for you to conclude, or understand my reason for choosing that size.

The front of the pigeon-holes must be made of a lattice door, with a hole in the middle large enough to allow one pigeon to pass at a time.

To this hole you must adapt by means of a hinge, nailed to the bottom of each pigeon-hole — a board, so as to be able to close the pigeon-holes when it is occupied.

The 70 cm. in depth must be divided into two equal parts by another railing forming a solid partition — half in wire netting and half in solid wood this must be hung exactly in the middle of the ceiling. It must be fixed on by means of two little hinges, so that it can be raised and lowered at will.

The round wires in the netting could be laid vertically.

A small nest must be placed behind the solid part of the partition, in which there must be a layer of earth or sand about 5 cm. deep. This solid half of the partition is of the highest importance and is commendable because the widower likes to go behind it to rest. It is also behind this part that you will put the female when you are preparing for mating. Then, during the competitions, particularly during the long distance races, it is in this spot that you will have a young pigeon taken over by the birds that like that place, and you can also set the eggs under them although the female be absent. The same applies to the pigeon-loft for the Natural, the set of holes are not to be placed on the floor ; for instructions look back in this chapter.

THE PIGEON-LOFT FOR THE SPEED COMPETITIONS

As the word « Speed » indicates, the pigeon-house allotted for this system must be combined, to serve in quite a different way from the other pigeon-houses.

First of all it is understood that the « speed » distance must not exceed more than 300 kilometers.

The fancier who likes these competitions must, if he wants to succeed, instal a pigeon-house adapted to this special branch of his hobby, neglecting nothing, his aim always being to increase the speed of his pigeons and not to allow one second to be lost through his own fault.

The loft or garret answers all the requirements of this branch of the hobby, as well as a special dovecote, only that the place must be big enough to be able to carry out everything that is necessary.

All the pigeon-holes must be attached to the same partition, between the pigeon-holes and the fence opposite, — there must be sufficient place to install the said pigeon-house, which takes the place rather of a garden where the inmates can have everything they need, and where they can live when they are not set at liberty.

This place must be about 3 m. wide, 3 m. long, and 2 m. high.

The pigeon-holes must be against the wall towards the outlet ; therefore in this manner there will be as many outlets as there are little lofts.

Each pair can dispose of a little dovecote of the following dimensions : — 75 cm. long, 60 cm. wide, and 60 cm. high.

Working exactly on these dimensions, you can have 12 little lofts in the same house, they must be raised 20 or 25 cm. from the bottom.

Always taking into consideration the home-coming influence, on the South or South-East side of the building you must bore holes in the partition, to communicate with each little pigeon-hole.

Those that are directly under the roof, could have a glass tile overhead to let the light come through.

Those situated underneath, against the wall must be lighted by a little window arranged at the beginning of the entrance.

These entrances must be 30 cm. high, 20 cm. wide.

The window fitted into this space, must be divided into four equal parts, three of which must have a pane of white glass, to let in light but to prevent the birds from seeing through it ; the fourth hole serves as a passage. Each entrance must have a landing board, 20 cm. wide and 30 cm. long. Do not forget that there must be plenty of air and sun coming in, just as in the big pigeon-houses.

These pigeons-holes, or dovecots must have a passage (or lobby) opposite the entrance, this is formed of a wire netting between the passage and the pigeon-holes, with space enough at the bottom for one pigeon at a time to pass out.

The; must be two or three bobs placed in this passage so that you can set the birds free or close them up, at will : at the bottom leading to the inside of the pigeon-house must be a wire netting in the form of a folding door, opening towards the outside of the pigeon-hole so as to be able to examine how they go in ; to catch them, and to attend to them, etc., etc.

The little corridor must give free access to the interior of the place. This will always serve as a passage for the pigeons to come in and go out.

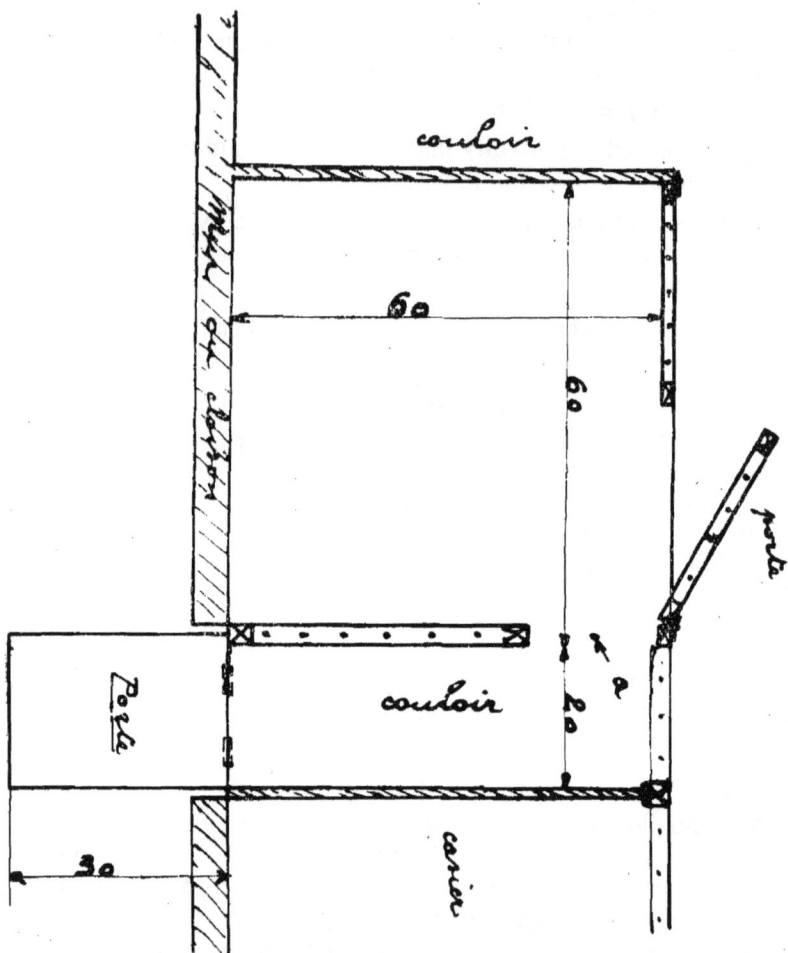

Loft for « speed » competitions

The pigeon-holes must be 60 cm. high, 60 cm. deep, and 53 cm. wide, plus 20 cm. for the width of the passage.

In so far as is possible fix the window so that two panes give light into the pigeon-hole ; the third one and the hole, lighting up and ventilating the corridor.

Into these pigeon-holes you must put two nests containing sand or dry earth 3 or 4 cm. deep.

Over these nests you will fix a board, 30 cm. long by 25 cm. wide, this is to procure place for the one that does not occupy the nest, to rest.

It can be closed outside by means of a little slide with panels. For those situated under the roof one « bob » allowing the birds to enter, but preventing their coming out, is sufficient.

In the part where the little pigeon-lofts are it is necessary to place a small board, for the pigeons that have no pigeon-holes to rest on, but on condition that they be 30 cm. large at least and placed at a good distance from each other.

In this spot which is a sort of yard or garden, they must always have grit, and also their food during the resting period.

They will take their bath there too. It must be well lighted, but the skylight must only be opened when all the other holes and the entrance are closed, so as not to cause a draught.

Another necessity of this system of the hobby is that the males must be set apart from the females and to go in or out they have to pass through the above mentioned corridor or passage.

This is to excite the jealousy of the males already mating, and designed for races.

The former only have a little board to rest on thus the others will become jelaous of their females and pigeon-holes.

The pigeons alloted for this role must also be birds of value, which in case of an accident, or loss of a racer can be taken out to replace that other one.

PIGEON LOFT FOR FEMALES

As is the case for all the other dovecotes, this one also must be constructed under hygienic conditions as stated in the preceding pages.

It must be placed as for as possible from the other lofts, because the males must neither hear nor see the birds.

It can quite well do without an outlet, but it must be where there is plenty of light and air, the front part being made in latticework.

If they can be percerved by the widowers during their exercise, put a black linen curtain in front, and draw it down during the daily exercise.

It is also preferable not to have any pigeon-holes inside, some little brackets placed on the right and left will do, the dimensions to be : — 15 cm. long by 15 cm. wide, so that only one bird can occupy it, if those are arranged in the wall the part touching it must be preserved from dampness by a board.

If, in spite of all your precautions, you find two females pairing, take one away from the loft and put her into a basket, for a day or two, then when you put her back be sure to take the other one out and give her the same treatment.

The females can play in the competitions like the males but in this case you must let them have exercise two or three times a week, and let them go into the pigeon-houses of the males after having about an hour's training.

Before letting them in be sure to take out all the males — except those that are already mating, that is, if they are not to take part in a copetition soon.

What is sometimes discouraging is that when the males and the widows race, on their return they arrive sometimes together and say on the top of the roof. To remedy this if is necessary to train them well, so as to obtain a speedy entrance.

GENERAL FACTS

When one wants to be successful in pigeon breeding it is not sufficient to possess pigeons of great value, one must also choose the sort of hobby that you want to carry on, and construct the pigeon-house accordingly, for the differences which exist between the different models have a special purpose. The great champions will take good care not to divulge their secrets, but if you pay a visit to any of them and that they let you see their installation you will be astonished to see all the little details, about which you would have never thought. As in every other sphere, a certain amount of knowledge built on certain principales is necessary for success.

You ought not to be satisfied with wiming prizes for one or two years, you must be able to continue to win success, and show that you are jealous of your knowledge concerning, pigeon breeding.

CHALET OR LITTLE HOUSE

Quite a number of fanciers prefer «chalets» to dovecotes in the attic, first because access is easier, and then the premises afford much better opportunity to practise the different systems of the hobby.

He who desires to build a chalet according to the conditions necessary for our sport should first of all think of the finding of the cardinal points then the material to employ (see page 122) in its construction, and also its position.

It should be according to the following dimensions : 3 m. wide, 2 m. high, and as many times 2 meters in length as there will be pigeon-houses.

Out of the 3 meters in width you must reserve one meter for the corridor behind the dovecots, and to install all the stuff necessary ; put blue glass in the doors and windows so that you will be able to see everything that is going on inside, without being obliged to go in.

The entrances to the dovecots must be on the south side if possible and must have all the accomodation required for the hobby you want to practise.

The skylights must be placed, one in each dovecot directly over the entrance, in this way you will have plenty of light coming in and at the same time all the bad smells and dust will be carried away.

The roof must be sloping a little ; and quite even ; it is also necessary that the glass in the skylights be a dull white ; you know why I recommend this. -:

The whole must be set up on wooden posts covered with zinc to prevent the cats, rats, etc., from getting in to the poultry.

The roof is to be of flat tiles with no ceiling nor lining of any sort.

If possible lay out a spot for the food in this part, cause a draught, and let as much sun and light in as possible to keep the grain always dry.

Each loft must have a seperate entrance except those for the pigeon for natural or partial-widowhood and for the young ones of same year, which must be arranged as I have already indicated.

The entrances must be made very high, one for each pigeon-loft, and the pigeon-holes must be situated on the right and left, never opposite.

If you practise widowhood, you could utilize the underneath part of the «Chalet» as an aviary for the females, or for the (sires) males, anyway be careful that cats or any other dangerous animals cannot get in.

These aviaries must be fenced on the North and West by a solid partition, on the South by a latticed one, then you must put up a shutter or a black linen blind, to be used during the hours when the widowers take their exercise.

Access to the Chalet must be by a stairs strongly made and in such a way as to prevent accidents and to leave no difficulty when you have to go to the passage ; it must have a railing too.

It is also well to make provision to protect the Chalet against all inclemencies.

Food

Many different sorts of grain are given to pigeons but all of them have not the same nutritive value ; very often it is damaged and unfit for consumption.

The principal reasons that grain, which, appears quite good is unfit for consumption, are :

1) The ground on which it has grown ;

2) The season in which it was reaped ;

3) The sort of weather;

4) The climate or part of the country;

5) The way in which certain grain has been cultivated;

6) The quality and quantity of manure used.

What every fancier must know is how to distinguish good grain from bad grain, and to judge which sort is most suitable for his birds.

He must also know what sort of mixture he must make, and to which class of birds he must give it, for instance, the breeders and the racers ; and the special ration to be given during the resting season.

Here I shall do my best to indicate the way to distinguish good grain from had, or the middling quality, and why not use the latter.

Musty grain brings on intestinal troubles, and sometimes may cause death by poisoning.

It is easy to know this grain as it gives off a bad smell, which can contaminate the other grain, even though both are not mixed together ; just by putting the latter into a vessel where the former had been, and which was not properly cleaned and aired.

The colour of damaged grain is often greenish but in the shops they, employ a sort of mechanical brush to clean this off, so that after the brushing it appears like fresh grain again, only that the taste and smell still remain, but either can be easily perceived.

Some grain contains impurities such a mustard seeds, ejections of cats, rats, mice, etc., and sometimes you get only the shells of the grain; the inside has been eaten by insects.

The odour of this grain is repulsive, and only very hungry pigeons will eat it.

Grain which has been reaped too soon, gets musty, it has a greenish colour, is smaller than the good seed and slips out of the hand very easily.

Whenever the seed is damaged or gnawed into you must not give it to the birds.

If the grain is effected with black rust, the pigeons will get sick if they eat it, and pine away.

When the seed is rotten it goes into powder and the smell which comes from it is like that of sulphuric acid — a dealy poison.

CHEMICAL COMPOSITION

In most of the seeds used you find the following composition :

```
Water ... ... ... ... ... ...      11.8 to 14.5 %
Digestive Albumin  ... ...       6.8 to 34.7 %
Sugar ... ... ... ... ... ...     10.2 to 72.7 %
Fats ... ... ... ... ... ...       0.3 to 40.4 %
Mineral matter ... ... ...        0.5 to  4.5 %
```

1 kilogram can give 2630 to 5131 calorie value.

1 kilogram : Starch varies from 59 to 130.

These are therefore strengthening, wholesome and natural foods, suiting the nature of the racing pigeon ; but of which one must have a good knowledge.

SPECIFIED VALUE OF FOOD
(FEEDING)

Feeding is of the foremost importance, as far as breeding and racing are concerned.

The fancier very often neglects the point of choosing proper seed; he pays too little attention to it and makes a great mistake, which causes him much trouble later.

When you feed your birds with inferior or damaged seed first of all you make them sick, and then they lose their force and you run the risk of ruining everything.

In the following pages I shall give you the details of several sorts of seeds used in pigeon-breeding.

MAIZE

Maize is an herbaceous plant, belonging to the family of the gramineous plants.

Many people think that it is only grown in America but it is sucessfuly grown in several countries. Is an excellent grain, and enters into the composition of the diet set out for the different systems of our hobby. It is rich in fats and sometimes contains hydrate of carbon, with wheat, oats, and rye. It contains less albumin, but the percentage of starch is greater than in any other corn. This seed is to be higly recommended, only its use must not be abused, or it will bring on a serious affection such as scurf, which is characterised by the epidermis becoming yellow, the body becomes covered with pimples, and the feathers being too dry fall off in quantities, etc.

As this grain is a choice foodstuff which the pigeons love, give just enough of it, but not too much; for too much of anythnig way cause trouble.

There are different kinds of maize, which grow in different countries, but I only know of one which gives satisfaction; it is the « cinquantino-plata » maize which is rather flat, bright red, yellowish and is often called « moyen », middle sized, maize.

There is also for sale maize known by the name of « Pignoletto », which I advise you not to use, for several reasons; this maize, if it has been picked out from the cinquantino-plata is not sound, it has not had the opportunity of coming to full maturity.

If it comes from the regions of the Danube, it has been reaped in wet weather and is therefore liable to get heated.

Be very cautions, for perhaps when buying it, it may seem better than the cinquantino, as it is specially prepared, but after two or three months it will heat, rot, and destroy the other grain that is mixed with it.

When you want to make your provision, take the cinquantino-plata; sort it yourself, and keep only the large grain (seeds), those are certainly very sound and ripe.

Above all be on your guard against maize with verderané.

Such a thing only effects this plant, it is a mushroom which, when people eat it causes a kind of scurf to come out.

BARLEY

Barley is nearly like wheat, it has little furrows or ridges, it is rich in fats but poor in albumin and is more digestible than maize.

Therefore I advise you not to use too much of it. Then again it has a disadvantage which can cause serious trouble, it ends in a thin dry point, which breaks like glass, and pricks like a needle, and may scratch the intertinal tube of the pigeon.

This grain is short when it has gone through a certain mechanical preparation to take away the sharp point, but no matter how well it is done there is always danger.

If the shell has been taken off in the process it loses some of its qualities, and after a short time becomes musty.

OATS

Oats has been cultivated for many years in Belgium it first came from Persia ; for our pigeons it should be considered only as a temporary diet, and should be given in the winter season to prepare the pigeons for moulting and then to bring them round to a good condition for training again.

Good oats is heavy, and is completely covered by a bright thin shell, without ridges, the fecula is white and abundant, and it has a very agreeable taste.

It must be giver during the same period and in the same proportions as wheat and rye.

This grain does not contain so much hydrate of carbon, albumin, or fats as the wheat and rye.

It should be given from the 15 th September until the mating season; pigeons of one year can get it until a week before their training begins.

Oats is not so often diseased as other corn. To be sure of procuring good oats always buy that called « bec de moineau » (sparrow's bill).

At the time of sorting the grain, before the sowing season, choose your grain; you have the opportunity at this moment of obtaining excellent grain for the pigeons. It must be a light golden colour and well rid of any beard.

WHEAT

Wheat is justly classed as the best cereal, seeing that it is the chief food of mankind.

This grain being rich in albumin, is very suitable for the upkeep of the whole system.

The great quantity of hydrate of carbon which it contains renders it equally suitable for birds that have hard work to do.

Wheat is a very precious grain, and contains a high percentage of fats ; it is one of the principal foodstuffs.

It grows in nearly every country, but the climate and weather have a great influence on this corn and the grain is not always sound.

To sum up its qualities should be :

1) To be free from black grains or germs; the grain becomes black through a disease called « charbon » which reduces the grain to powder. It is a deadly poison, and during the threshing this powder flies on to the other grain and poisons the lot.

2) The grain which contains germs has been reaped during bad weather;

3) The wheat grain must be small, russet or reddish in colour, and smooth.

When the seeds have ridges it shows that it suffered during the growth, and it is better not to use it.

When you chew wheat, you have a sweet tast in the mouth ; a bitter taste denotes that the corn was reaped during bad weaher.

It must have no smell and be quite free from ejections of rats or mice.

Of all the different sorts; the small russet grain is the best, it is the « Ièverson » seed imported into Belgium, and difficult to find, but in default of procuring this you can get exotic wheat called « Manitoba » at a miller's; this also as a good nourishing food for the pigeons.

Newly reaped wheat, favours the moult.

RYE

Many fanciers will not feed their pigeons with rye, as the birds do not like it.

However this cereal must at a certion season, be included in the rations.

This grain is distributed in Winter at the same time and in the same proportions as oats.

Rye keeps the pigeons thin, and the result is good for the down.

It is cultivated nearly everywhere and grows chiefly on poor and barren lands, like barley and wheat, it can be effected with diseases such as, black rust and ergot.

Ergot is of a dark purple colour, and upsets the system of the pigeons that eat grain effected with it.

The principal point about this grain when you are buying it is to see that it is quite smooth and free from spots.

RICE

Rice is an excellent food, refreshing and at the same time nourishing (a complete food). It is very good for young pigeons and for those which must take part in the long distance races.

Rice is rather a preventative of diarrhoea and small quantities should be constantly put into the pigeon house of the producers and breeders, and also in the loft where the pigeons of under twelve mouths are kept.

A little rice should also be given to the pigeons in the evening feed, during two or three days before being put into the baskets; that is for those that race a distance of over 500 kilometers.

The husk must be taken off the rice.

THE DARI

The dari is also an excellent grain, which the pigeons like very much, like the rice it is a prenvtative of diarrhoea too.

The plant grows to about 4 or 5 meters high and is cultivated in hot countries.

There are several sorts of dari for sale, red, grey, yellow, etc., etc. But the very best quality that exists is pure white.

For speed comptitions, as well as for the natural and partial widowhood this grain is excellent.

The dari should be given in the morning feed.

BUCK-WHEAT

Buch wheat belongs to the family of the polygones. It is cultivated or grown in poor lands. It is splendid food for sick pigeons or those that are convalescent; rich in fats and more digestible than wheat.

Generally, young pigeons love it, and it renders their flesh hard and firm without making them too strong, and fat.

It has an advantage over other grain, as it is a great help when the fancier percieves that the moult is late amongst his group of pigeons.

Buck-wheat makes the feathers smooth and silky with plenty of beard, and barbet.

I should advise you to use it at the end of the season, especially when there is a tendance to a late or irregular moult, and also for the pigeons not yet a year old.

MILLET

The millet is mealy and rather « excitable ».

It should be included in the diet or ration, after the races, as it is a tonic.

Like buck-wheat it can be given during convalescence or to sick pigeons.

There exists flat millet and round millet, then again there are many different qualities.

Preference should be given to the white smooth millet, which is a good heating food.

Do not use the yellow or the red millet.

The flat millet makes the pigeons hardy, but it is dangerous to include it as part of the food.

From my experience, I have seen that birds which had eaten flat millet often had scratches on the interior of the gizzard.

Millet has a splendid effect on the feathers it makes them silky and contracts the tissues.

LENTILS

Lentils are an excellent foodstuff, but highly nitrogenized.

Besides the nitrogen, it also contains a considerable quantity of fats which are indigestible — therefore injurious — especially to the birds which are allotted for speed competitions.

Nearly every bird-seed merchant sells lentils ; very few have sound grain, even few of them know it.

There are also many different varieties, that which is most sought after and preferred, is two years old and of a brownish colour, like bricks; the inside is orange colour, and is of the average shape or size.

Young lentils must not be used, they are easily known as they are of a yellowish — or yellowish-green colour.

PEAS

Nearly all the different varieties of peas possess the same qualities and the same richness, therefore they all are quite suitable.

Only be cautions when you make your provision that they have not a musty taste or smell, and that they are not « bruchès ».

There exist many varieties of peas; branch-peas, dwarf-peas, green peas ; yellow peas ; English peas ; grey peas, etc., each sort is known by its colour.

VETCH

This is a vegetable belonging to the papilionaceous family; cultivated in different countries as fodder, and for the grain.

There are several varieties, but two are well known, the indian vetch and the exotic vetch.

The exotic vetch which is bigger than that of this country (Belgium) generally comes from Germany.

Indian vetch is small and of a greyish colour it is preferable to all others.

It is excellent foodstuff and can take the place of peas and beans in your rations.

For several years it has been difficult to procure a very good quality, and it is my opinion there exists only one sort upon which you can rely, it is the small, grey vetch grown in this country (Belgium).

The bright, or black vetch, coming from the sorting must be discarded. These sorts are generally a kind of mixture; or have been specially prepared for sale.

They often contain what is known as blith (niellé) a deadly poison; after being brushed and prepared for sale this cannot be perceived by the buyer.

To day, we have arrived at such a degree of perfection that although the seed is full of insects, unfit for sale, and musty it can be made to look clean, bright, and attractive, but still be full of germs.

There is nothing to be done; there still remains a bad taste, and on breaking the seeds you will see traces of little grey lines.

Later on, this will cause much trouble. If you like to feed your pigeons on vetch, buy the grey sort, chew a few grains, they should taste like nuts, and the inside should be white.

Grind a handful in a piece of white cloth, the good quality will leave no traces, make your provision, spread it out on a floor to air, or on some place well protected from rats, mice, etc.

Here I must tell you : I have seen black shiny vetch, which from all appearances was of a good quality, but before ordering the provision the person put a little of it into warm water and the result was that the black vetch became grey, the black was just a colouring which had been given to several different varieties to form one sort.

BEANS

Beans being included in large quantities in the rations must be well examined. Like all other grain, there are many different varieties, which vary according to the conditions and places in which they have been grown.

However real good beans are rare.

The best are the English beans, a year old, rather large, round, without ridges, a light brown colour, and the inside a light yellow, having a nice taste, not at all bitter; smooth and clean.

The dutch beans are well known, only having heen cultivated in a damp country, it is difficult for them to be unreprochable, they are smaller than beans grown in other countries.

Those from France have no flavour and are often perforated by insects. Those from Luxembourgh are so sour that you cannot use them.

HEMP

Hemp seed is very rich in fats, and especially in albumen, there should be about 50 % of it used in the rations given to regain strength in certain systems of the hobby, especially for the speed competitions.

Of all the different varieties for sale, that which gives most satisfaction is the very large hemp seed from Chili.

It is easily known, because one can perceive some grains here and there which are not altogether ripe.

The seeds are small and of a greenish colour. Hemp is also cultivated in Belgium, the seeds are smaller and blacker, but are of a very good sort.

Always choose hemp-seed, with a large grain it has an agreeable flavour, and tastes like nuts.

The small grain is not quite ripe, so it must be discarded.

Before using this seed take off all dust, and pick out any seeds that are perforated, or which may be injurious to the birds.

A good way to do this is to steep the seed in a bucket of water over night.

In so doing any straw, dust, or bad grains will float on the top of the water, and then you can easily take them off. The water must be renewed until it remains quite clear, then the grain is well cleaned. Dry immediately in the sun, or otherwise; stir it from time to time to prevent fermentation.

Fanciers who give the grain to the pigeons just as it has come from the shop make a great mistake. To be convinced of the fact take about one pound, wash it and you will see the dirt it contains, and which you would give your precious poultry without doubting you were doing wrong ; you will certainly be astonished.

RAPE-SEED

This seed should be included in the composition of the mixture for the moult, and also in the rations given during the racing season, and to the birds for natural and partial widowhood, and for those set out for total widowhood too.

This grain is the richest in fats, it prevents the pigeon from getting thirsty, and has a splendid effect on the feathers.

The pure black rape-seed, quite sound is very difficult to get as it is often adulterated, so as to hide the defects.

This is how different kinds of grain. appear to have a lovely black colour, chiefly rutabaga when sold for the best rape-seed.

If some seeds have a bad taste they are put into a special mechanical brush and in a very short time through this operation the superfical defects are taken away.

Then the grain is passed into another machine where it is soaked in oil, this gives it a shiny black colour.

The really good black rape is rather of a dark grey colour not dyed, nor oiled, nor brushed and contains no middle sized red seeds.

It becomes ripe about the end of July or August. It is grown in Belgium. The black rape seed has a nice taste, and leaves the impression that it is greasy, and does not prick the tongue.

RED RAPE SEED

This grain comes exclusively from foreign lands, the genuine grain is very difficult to find and especially sound stuff.

The only sort which is considered of a good quality is of brick colour, it is smaller than the black rape seed.

The salesman who offers you large red seeds saying it is the best quality, only offers you mustard or turmp seeds, just taste them and you will find that they burn the tongue immediately, naturally if you give this to your pigeons they will all become ill, making it impossible to class them for the competitions.

Owing to this, I have seen fanciers who used to be successful, during a whole season, have their names figure on the list only once.

RED CABLAGE SEED

Many pigeon breeders deplore the high price of red cabbage seed — (100 f. to 120 francs the kilo — note that with 250 grms. you can easily have six pigeons race the whole country). (Rations : pages 196 to 199) and wonder with what they can replace it.

This is a heat giving food, but it can be replaced by fenugreek, (which is not quite so heating) or by a concoction of celery; 2 soup spoonfuls of the dried leaves (or stems) to a litre of water.

This will have an effect after a short diet without of course neglecting the other prescriptions in the course.

LINSEED

Linseed ought to be given every day during the resting season, and also on the home-coming. It is most essential, on account of its special properties which act on the blood and on the feathers.

As a laxative, give only a small quantity. In greater quantities it is an excellent purgative.

There are two kinds of linseed, that which grows at home, and that which comes from Morrocco.

The seed from this country (Belgium) is small and ferments easily, it is richer in fats than that coming from Morrocco, but the latter is preferable, as the pigeons can pick it up more easily.

In order to have a suitable diet for the pigeons there are other products to which you must devote your attention : they are, toasted bread, new bread, sugar, mineral matter, salt, and greenstuff.

« COMBAT » BREAD

It must be made as follows :

Cut 250 grms, of new bread into slices of about 5 or 6 $^m/^m$, thick and cut the lot into little cubes about the size of grains of maize, put into the oven to dry, every time you need some, first put a piece of butter about the size of a nut on the frying pan on the fire to brown, then place your little cubes into it and stir them until they are well toasted.

Leave them to cool, then shake some sugar over them (about two lumps ground up) add the yokes of 6 new-laid eggs, stir in and beat the lot together until it becomes assimilable, and the little cubes become impregnated, after this, leave the whole to dry and your « combat » bread is ready.

It is not necessary to insist upon the quality of home-made stuff. This should be given to the birds taking part in the speed competitions as well as to those for long and short distances.

Toast should also be cut into little cubes like the bread for the « Combat » food, and should be given as will be indicated later on.

NEW BREAD

New bread should be given every morning to the cocks and the hens during the breeding season.

The youngsters should receive new bread from the time of their weaning until they have moulted four or five wings. From time to time they should also get toast, this latter should be included in the feed when they have to take part in the competitions.

Bread is simply cooked grain, with the husk taken off. It is especially suitable when the down must be shed.

It is more digestible than the raw grain, new bread and toast are to be highly recommended and should be included as part of the rations during all seasons.

It is a most excellent foodstuff for the young ones, as well as for the fathers and mothers.

Experience has proved this.

Toast should not be crumbled or ground up, there is too much waste and this collects rats, and mice.

SUGAR

Sugar is a complete foodstuff, totally substantial.

It suits therefore as part of the rations for the racing birds, especially those participating in long distance races.

We have seen in the first book that sugar is necessary to the pigeon, and that it even makes sugar itself from the food supplied to it, It accumulates a reserve for later consumption.

If you desire a development of the whole system it is necessary to give sugary foods, which is without doubt the choicest food for our birds as it increases muscular energy and resistance.

The proof that sugar assimilates so well in the system is that it leaves no trace in the droppings.

It has a good effect on the respiration, renders it slower and regular, reduces breathlessness, prevents fatigue, makes the muscles stronger acts on the circulation of the blood, stimulates the heart, keeps off thirst, prevents perspiration, helps the digestion, and strengthens the bones.

Sugar must be given in little pieces, the size of a grain of maize. There is a sort of pearled sugar sold — n^r 2 and n^r 3 which suits very well, as the pigeons can easily pick it up, after a little practice.

At the end of the training period, the birds should receive two little pieces every evening, increasing this number as the time approaches for the long distance races.

Sugar must also be given in the « bread for the combat », in the drinking water when the racers return, as well as in the drinking water for the feathers and mothers and the youngsters not yet a year, during the moult, and always in tea, whenever it is given to them.

MINERAL MATTER
GRIT

There must be a covered bin placed at the disposition of all the pigeons, containing a mixture of different kinds of gravel and grit, such as washed river sand, old mortar, offals from bricks, flint, charcoal, granulated bones ; then grind the lot together mix with egg shells, add 100 grms, of sulphor, (washed) a cupful of salt, and one hundred grammes of ground aniseed ; pigeons are very fond of this mixture ; which aids the constitution, and enables them to surmount great difficulties.

The component parts act favourably on the development of the skeleton.

SALT

Salt is a condiment which every fancier should make sure to have in a little bin in the loft.

The pigeon not being very fond of it, will only take a little of it from time to time when its system needs it.

If you deprive your birds of this condiment, they will lose their appetite, the skin will become soft, the muscles weak; thus they will lose strength and not be able to fly the distance for which they are entered.

Salt is known to be a nutriment of the greatest necessity. It is found in certain lakes, in the sea, and also in rocks; this last mentioned you can buy at the seed merchants it is called rocksalt.

Some fanciers like rock-salt; however, you can obtain a good and cheap rock by simply taking as follows :

1 kilo of table salt, put it into a cloth, damp it with water or steam, then put it to dry in the oven or on the mantelpiece and you will have the finest and best rock existing.

Salt is highly recommended, it is an excellent for the digestion gives the pigeons an appetite, and clears the bronchial tubes.

GREEN-STUFF

Greenstuff is very necessary for the pigeons. To prove this you can examine carefully a bird that goes into the fields, — when the gizzard is full you will find much greenstuff, grit, snails, insects and but little grain there.

If sugar and salt are highly recommended for the growth of the bones, greenstuff is as necessary for the blood.

In general, any greenstuff is good; but the best kinds are water-cress; endives, chicoree, lettuce, In default of these you can give green cabbage, or even, a scraped beetroot.

Grèènstuff, when very tender, possesses refreshing and wholesome properties, and has a most beneficial effect on the birds. It contains a high percentage of phosporous acid. iron, and lime, all of which help the constitution and strengthen the blood.

Greenstuff also contains a high percentage of vitamines, substances which are not yet thoroughly known.

I advise you to give as much chopped greenstuff as possible, the whole year, to the young birds, the producers and breeders; and also to the racers the day after their return from a race. This is to avoid if possible catching any very bad disease.

CHOOSING THE GRAIN

When making their provision, how few fanciers trouble to choose the best quality.

It is however necessary to know how to choose between good and bad grain.

During my several years experience and at the sittings held for sorting the birds, have I not had the opportnuity to judge that pigeons which were apt to be first could but seldom be classed owing to having been fed with damaged grain.

Therefore only take the best quality for which you will be highly compensated.

Good grain can be known as follows :

1) It must be smooth and have no little lines or ridges in the shell ;

2) In a large quantity of grain all the seeds must be of the same size, and form;

3) The inside must be of a light colour, and without any streaks or spots, it must have an agreeable flavour ;

4) It must be free from dirt, or ejections of rats, mice, cats, etc., etc.;

5) The grain should not be weevilled, spurred, perforated, etc., etc. Any grain which has a bitter or sour taste must be thrown away.

The grain ought to be dry. By thrusting the hand into the bag, you can feel if it is damp or warm or if it contains dust.

Put a handful of the grain into a white cloth, squeeze tightly and rub together, it must leave no trace of colouring; if you put a few grains into the mouth and they change colour, be on your guard.

(t) = Take a handful of any kind of grain, let it fall gently into the bag, the seeds must have a ringing sound as they touch each other.

Before making your provision, ask for a sample of each sort, then you can make your choice and order your provision. When you receive the grain do not leave it in bags. Make a little safe with boards on the sides, the top and bottom being in tin or some other metal, arrange several little cases or drawers in which to put the different sorts of grain.

It must not all be mixed together, only when it is needed. Then at least once or twice a week you must stir it up so as to let it dry and to let the dust fall out. The box containing the mixed grain ready to give as foodstuff, must also have a tin bottom; before giving the mixed grain to the pigeons it must be screened.

DRINKS

Water is strictly necessary for pigeons, it is the natural drink of all human beings, animals, and plants, there is plain drinking water, mineral water, and medicinal waters.

The body of the pigeon contains about 50 % of water, the percentage reduces as the pigeons get older.

The water which you give your pigeons must be drinkable — clear, fresh, and without a smell) — put in into a stoneware vessel and let it stand for some time in a corner of the pigeon-loft before giving it to the birds to drink, so that it will become the same temperature as the air in the pigeon-house.

If the water is too cold it can bring on, typhus, cholera, diphteria, thrush, croup, diarrhoea, pneumonia, enteritis, etc.

When your pigeons are suffering from anaemia, or are run down after racing or rearing the young ones, it is a very good thing to give them water with iron in it; you can just place an old piece of iron in the trough.

The sort of water which is given to the pigeons is of the greatest importance, it is therefore necessary to know something about it. First the drinking vessels must be quite clean; the water, pure, wholesome, exempt from all microbes and carried to the pigeon-loft in clean vessels.

With regard to hygiene, — water plays a very important part — as it can also be one of the chief means of developing microbes.

Diseases are also spread through impure water. The sort of water given varies according to the season — the Winter season, the racing season, and the moulting season.

WINTER SEASON

From the first of January to the first of march for any sort of loft, the drink given to the pigeons during this season is to be prepared as follows :

Take a litre of water and make an concoction of some quinquina shells and some sarsaparilla — a soup-spoonful of each ; leave to boil for 5 minutes, add some pure honey, a soup spoonful — in default of this — five or six lumps of sugar — set aside to cool, and then give it to all the group.

This infusion, given during the Winter season brings on a thorough shedding of the down.

It should be given twice a week; on the other days give fresh water to which has been added a few drops of tincture of iodine.

RACING SEASON

Honey should be added to the drinking water — a soup-spoonful to a litre of water — from the beginning of the training until the races are over of course the honey must be melted first in half a cupful of (warm) hot water.

On the other days let a few drops of iodine fall into the drinking vessel.

This drink helps the bronchial tubes, and takes away any inflammation caused by certain efforts made on the way.

THE MOULTING SEASON

The water given when the stages are over, up to the 31st december must contain the following products :

At the end of the week — during two days, give water containing iodine, (about 4 drops to a litre of water) the next two days, add five or six lumps of sugar or a spoon-ful of honey, as already indicated. This is to strengthen the bones and clear the bronchial tubes.

The three following days I should give an infusion of plantain prepared in this way :

To a litre of boiling water add 25 grammes of plantain; leaves, stems roots and all. — it must be quite dry and free from dust, or any waste matter.

Boil during 10 minutes, add 5 or 6 lumps of sugar so as to take away the bitter taste, then strain through a clean cloth, leave to cool before giving to the birds to drink.

This makes hhe feathers firm and abundant, helps the blood and banishes dandruff. You will see the effect yourself.

MEALS

The racing pigeons must have two meals a day throughout the year. The first between nine and ten in the morning, and the second between 5 o'clock and 6 o'clock p. m.

The morning meal is at about the time when the pigeons generally return from the races.

The feeding hours must not be regular for the males and females during the breeding season.

As early as possible in the morning give one meal and another during the day whenever it is necessary ; then a good meal as late as possible at night, so that the young ones can have their fill.

For the racing birds the rations must be given after each flight, from the beginning. of the training to the end of the stage.

Each pigeon should have a little earthen-ware trough in its pigeon-hole, where it gets its own ration : thus you are sure of : 1st) the quality of the ration ; 2nd) the nature or sort ; 3rd) the quantity.

If it is necessary to change the diet, to suit the different competitions, or during the resting period; it must be done with great prudence.

A sudden change is very dangerous, as it can bring on diarrhoea or inflamation of the bowels therefore it must be done gradually.

RATIONS

The rations are composed of different sorts of grain, according to the period and the role which the pigeons have to accomplish e. i. breeding ; racing ; etc., etc.

There are therefore three distinct periods; and also several distinct pigeon-houses but the birds in each, must on no account get the same sort of grain nor the same diet.

DIFFERENT PIGEON-HOUSES

1) Breeding loft ;
2) Loft for natural and partial widowhood ;
3) Loft for complete widowhood;
4) Loft for speed competitions ;
5) Loft for the young birds.

1°) LOFT FOR BREEDING

During the winter season, from 1st January to the 15th February, the ration must be given half in the morning and the other half in the evening; about 25 grammes a day for each pigeon the composition includes 30 % wheat; 30 % rye; 30 % oats; 10 % linseed. Plenty of chopped greenstuff.

From the 15 February to the 1st March the mating period, by degrees you must add large grain (seeds) to this mixture : As follows :

From the 1st March to the 1st September the food must not be limited and should be composed of the following : horse-heans 25 %; peas 20 %; wheat 20 %; maize 10 %; rice 5 % ; linseed 5 % ; new bread 5 % ; candied sugar 5 %; chopped greenstuff rape-seed 5 %.

Between the 1st September and the 31st December, give plenty of the following mixture : horse-heans 35 %; peas 20 %; maize 10 %; wheat 10 %; buckwheat 5 %; rape-seed or senna 10 %; hemp seed 5 %; linseed 5 %; and from time to time add new bread with a little cod-liver oil, on it.

Once a week during all seasons, you should shake a teaspoonful of flour of brimstone over the food.

2°) LOFT FOR NATURAL AND PARTIAL WIDOWHOOD

During the Winter season — between the 15th December and the 1st February, the feed should be given in two parts — half in the morning between 9 and 10 o'clock — the other half in the evening between 5 o'clock and 6 o'clock : about 25 grammes a day to each bird, of the following mixture. (short) oats 50 %; rye 25 %; wheat 12 ½ %; linseed 12 1/2 % ; Plenty of chopped greenstuff.

Between the 1st of February and the 1st March, same food without increasing the rations. Only add a little wet brown bread and sprouting oats. Withdraw the (short) oats and the rye.

From the 1st of February until the races are over, the food should be given always after the flight, in little earthenware troughs which should be placed in the pigeonhole of each bird, entered for a competition; the other pigeons should receive the same food as is given in the evening.

The following mixture is for the racers : Chili hempseed 25 %; rape seed 25 %; white « dari » 25 %; toast 10 %; round millet 5 %; flat millet 5 %, red cabbage seeds 5 %.

For the evening feed : maize 50 %; horse beans 30 %; ajra peas 10 %; (wild peas) wheat 10 %, over a teaspoonful of « combat » bread and a lump of sugar about the size of a grain of maize : During all the racing season quantity of the rations should be about 20 to 25 grammes a day for each pigeon.

Every Monday or the day after the races the morning feed must be composed of pure linseed and chopped greenstuff.

You must make an exception for the birds taking part in competitions that same day or the following one.

From the 1ˢᵗ September to the 15ᵗʰ December, give plenty
of the following mixture :

In the morning : wheat 25 %; rape seed 25 %; hemp
seed 10 %; buckWheat 10 %; flat millet 10 %; linseed
10 % ; granulated sugar 5 % ; fresh bread 5 % ; to which
you have added a little drop of cod liver oil, to soften it.

In order that the pigeons will pick up all the sugar, put
the mixture into a tin box containing a little food, add a
handful of sugar mix all together until it sticks to the grains.

Before making this little preparation, be careful that he
grain employed is quite clean and do not forget the chopped
greenstuff.

In the evening : new horsebeans 50 %; « jara » peas or
wild peas 30 %; maize 10 %; new wheat 10 %.

3°) LOFT FOR COMPLETE WIDOWHOOD

During the Winter Season, between the 1ˢᵗ January and
the 1ˢᵗ March the rations must be given part in the morning
between 9 o'clock and 10 o'clock, and the second part between 5 o'clock and 6 o'clock in the evening — 25 grammes
a day to each pigeon.

The mixture must be composed of the following ingredients : short oats 50 %; rye 25 %; wheat 12 ½ % and
linseed 12 ½ %.

Between the 1ˢᵗ March and the end of the racing season,
the birds should be fed after their exercise, and the feed
should be given in the pigeon-holes, at the time when the
pigeons generally return to their lofts. — often between
9.30 a. m. and 10 a. m.

The ingredients are : Chili hempseed 25 %; rape seed
25 %; white « dari » 15 %; round millet 10 %; flat millet
10 %; toast 10 %; red cabbage 5 %.

In the evening between 5 o'clock and 6 o'clock, after the flight, you should give : a mixture of maize 50 %; horse beans 25 % ; wheat 12 1/2 % ; peas 12 1/2 % ; with this a leaspoonful of « combat bread ».

During the races the rations should weigh about 25 grammes, slightly increase this quantity according to the length or distance of the races. Those birds having to take part in long distance races should get two pieces of sugar about the size af a grain of maize.

The first meal of the day after the race should consist of pure linseed and chopped greenstuff.

Between the 1st September and 1st January plenty of the following mixture should be given ; morning feed : wheat 25 %; hemp 10 %; backwheat 10 %; flat millet 10 %; linseed 10 %; new bread 10 % to which you have added a little cod liver oil mixed with emulsion.

Do not fail to give greenstuff every day during this period.

In the evening : new horse beans 50 %; peas 25 %; maize 12 ½ %; wheat 12 ½ %.

4°) SPEED PIGEON LOFT

During the Winter season between 1st December and 1st February the same rations should be given as before that period, and should consist of : short oats 50 % ; rye 25 % ; wheat 12 1/2 % ; linseed 12 1/2 %.

You should give 25 grammes a day to each pigeon, and also plenty of greenstuff.

From the 1st of February until the training begins you must give the same quantity but you must gradually introduce the following mixture into the feed.

From the beginning of the training season until the races are over the morning feed should consist of : hemp 50 %; toast 25 %; white « dari » 20 %; and 5 % of candied sugar. Never give 25 grammes.

For the evening give ; maize 75 % « combat » bread 20 %; red cabbage seed 5 %.

The day after coming home from the races, the feed must only consist of pure linseed and chopped greenstuff.

Between the 1st September and 1st December give plenty of the following mixture : wheat 25 %; rape-seed 25 %; hemp 20 %; buckwheat 10 %; linseed 10 %; flat millet 5 %; sugar candy 5 %; and 5 % of bread softened by a little cod liver oil mixed with emulsion. Plenty of chopped greenstuff.

In the evening give : horse beans 50 %; peas 25 %; maize 12 ½ %; and wheat 12 ½ %.

5°) THE YOUNG PIGEONS

The young pigeons need plenty of good nourishing food : it must consist of : horse beans 20 %; peas 20 %; wheat 20 %; maize 20 %; new bread 10 %; buckwheat 5 %; candied sugar 5 %.

You must also give them, from time to time a handful of fine grain consisting of rice, hemp seed and some linseed. linseed.

In order to teach the young birds to pick up the grains properly you must put two or three hungry old birds into the pigeon-loft with them to induce them to eat.

Every evening you must examine the birds one by one, and feel their crops to see if they are full.

The birds should look bright and healthy.

When a pigeon droops, or looks to be out of form, it is better to lose no time but to make a few little balls of new bread, hemp seed and sugar. Put this into the bill, then make the bird drink, next day feel the crop which ought to be empty.

The very young pigeons should be carefully watched because sometimes they do not know where the fountain is, and then for want of drink they become weak and will pine away in a few days.

———

THE CARE TO BE GIVEN AFTER
THE RACES

After coming home from the races you should watch the birds very carefully. Do not do like some fanciers do once the competitions are over, walk around talking to this person and that one about the race, and all this time neglecting the poor tired birds that have been closed up for several day in baskets ; have flown for hours, sometimes the whole day in the rain and cold, or in the scorching sun.

They have used up much of their heating power and given out a great deal of their strength, in so doing the blood produces a poison which goes through the whole system.

It is therefore ncessary instead of losing interest, to pay great attention to the birds and to give light digestive food the first day ; the next day, a good bath, remembering to give something to purify the blood.

1) Nourishment : This ration, about 10 grammes should be given in the pigeon-holes so that they can eat it at their ease.

It should consist of equal parts of toast, wheat, linseed ; the same ration again in the evening.

The day after, on principal they ought to get pure linseed and chopped greenstuff.

The following day, in the evening you can begin to give the same ration and mixture that I have already described for each pigeon-loft.

As to the drinks; on returning from the races the drinking vessels should be empty so that the birds will not drink cold water which is very bad.

Although there be no water, the vessel should be in its old place. Half an hour after the home coming give them water which has been boiled and into which you have put a little honey or sugar. This little bit of advice if carried out will help you to avoid much serious trouble.

THE FIELDS

The time to go into the fields begins in August and lasts till October, or even longer according to the weather, the pigeons find all kinds of food they need in the fields. It is not always easy for those in towns to acquire the habit of procuring their own food in the country. In order to do this it is necessary not to give any food for several days — a thing which I am not in favour of at all for during the moult the pigeons need penty of nourishment, and must be in no way neglected.

For many years agriculture has been making progress, at all times of the year agriculturers use large quantities of chemical manures which are deadly poisons. And those highly perfectioned machines now used, do not leave much grain after the reaping, thus the birds cannot find sufficient food at this critical moment, it is therefore necessary that they should still receive food and greenstuff in the pigeon-houses.

One disadvantage of this method is that when the pigeons have the habit of running in the fields, they will go there too often when they are being trained or when taking their daily exercise, during the races, if they have any little difficulties they will return again and thus injure their health, and their coming into form.

Like that they can also be a prey to hawks, cats, or other birds of prey, and also get caught by inattentive huntsmen.

Also when a pigeon exposes itself to danger or « falls to pieces » as some fanciers say, or when it is caught in a storm, at a distance from the pigeon-loft it is impossible for him to return and is exposed to all its enemies.

Is it not easier to give the pigeons liberty and to supply them with everything that is necessary : greenstuff, condiments, etc., as I have already explained to you in the management of pigeon-houses during the Autumn season.

Thus you will avoid much trouble and serious losses.

IMPORTANT ADVICE

Pigeon fanciers do not pay enough attention to feeding their pigeons, they very often neglect this point although it is so important.

It can be said even to be the capital point to succees in this sport.

Very often people consider this question as secondary although they expect the pigeons to behave well during the races ; and to produce fine young birds without employing the means necessary for that purpose.

Naturally a pigeon that is not healthy or that has not the qualities necessary for a good racer ; can have the very best diet, be bred by the greatest master and it will never be a good bird for the sport.

But there are many pigeons according to the racing results which are only ordinary, or even bad racers, just because their diet has been neglected.

The cocks also lose their good qualities from one year to another for this same reason.

It is very sad to see every year as I do during the meetings, the number of pigeons that have not properly moulted or that have some disease of the blood all resulting from bad feeding, indeed, in such cases fanciers pay dearly for the consequences.

Fanciers who do not nourish their birds properly i. e. give bad grain, or lose interest in the pigeons once the races are over commit a crime for it is just then, that they require extra care.

Those who leave the pigeons to do for themselves will surely meet with disappointments, for the fields are so well raked up that sometines it is difficult to find a single grain.

The fancier who is anxious about his pigeons and who loves them, and aims at success feeds them properly during the moult, and he especially chooses the best grain as it is during this period that they prepare for the next campaign.

is it not therefore from the end of August until the middle of december that they must receive extra care, and plenty of the best new grain to help the new feathers to grow.

The bird that has slender, dry, brittle, or stiff feathers cannot, be classed, and will not even be able to produce good young pigeons for future compaigns.

The moulting season is considered to be between the the racing stages and the following year at the beginning of the competitions ; this period is a most important one if your group is to be kept in good condition.

The pigeon that has had a good moult and is in good health will have no difficulty in coming into form and will be classed in the first rank and at the head of the prize-list once the time of the competitions is announced.

The rations or diet which I have recommended for the racers, or for the pigeon allotted for any system of this hobby will be of no use whatever for pigeons which have not had a proper moult.

It is therefore for you, dear reader, to follow the good advoice which I have given above for each system of the sport.

The distribution of the food during the moulting season must be done with care, calculate carefully the quantities of grain you put into the mixture, and be sure te give plenty to eat.

I also attach great importance to the distribution of greenstuff which must never be forgotten all through this period.

Greenstuff contains phosphoric acid, iron, lime, it is refreshing, and purifies the whole system, it also contains many vitamines which are necessary for all living creatures.

Give water-cress, cabbage, salad, and all kinds of vegetables except acidulous stuff, such as sorrel. Give green stuff all the year, to every pigeon except to the racers during the races, when they must only receive green food the day after their home-coming.

HOME COMING INSTINCT

This subject is as old as the pigeon itself, and nobody to this day can claim the right to have made us any clearer on the point.

All the great master breeders tried to explain it but I think not one has been positive. Not one could give ample proof to confirm his writings.

Some of them lay it down to the memory, but in my opinion the memory of the pigeon is not superior to that of man. If somebody conducted you about 100 kilometers away from where you live — in a closed car and them set you at liberty in a country where you could get no information as to your where-abouts do you think you would be able to find your way home without knowing the place, and even if at a certain moment you were allowed to cast a glance, do you think you would be likely to recognize a certain point or landmark in the distance ; man, most intelligent of all creatures, can you answer this question ?

The pigeon sees nothing on the way but things which do not interest it, and still it can find its way home.

Others attribue this faculty of finding the way home to the sight. No doubt they have forgotten their geography, for the earth being round, to be able to see at 24 kilometers distance the pigeon should have to fly 60 m. high at 48 kilometers it is necessary to be 240 m. high, and so on, to arrive at 400 km. it should go up to be height of 15.030 m. then, we know that when the aviators take up pigeons and set them at liberty at a height of 800 m. they let themselves fall down vertically they do not fly, this is, therefore a proof that the pigeons cannot fly at such a height.

That the pigeon has a sharp sense of sight, I admit, but it is not only the view that can indicate the direction it must take, it has also been noticed that a pigeon set at liberty 100 leagues from its loft immediately takes the right direction and continues on its way without hesitation.

Other fanciers have even said that the home-coming was due to the sense of smell.

However the pigeon has not a very developed sense of smell, and I believe this theory is not the right one either.

The sense of sight has certainly a lot to do, but in the long distance races, in coming home during dark, cloudy, misty weather, necessarily there must be a factor superior to all others which leads the pigeon through plains, valleys, and mountains, this factor is, I think, the hearing.

As soon as the pigeons are let loose, there are some which take a determined direction, others fly up to heights varying from 5 to 60 metres then at a certain moment, they also take a certain direction and fly straight on to a fixed point — their loft.

One would think that while they were flying round and round they drew up a plan of the place or locality, made a land mark and then coming to a decision, off they went in the direction which they supposed to be the right one.

In my opinion to explain the real means by which the pigeon directs its home-coming we must have resource to science — the atmosphere is electrified or contains electricity not only during a thunder storm but at all other times.

When we have a nice blue sky the electricity in the air is positive, it is perceptible only at a certain height from the ground according to the determined altitude at which one is placed.

The layers of electricity also vary in intensity according to the time of the day or the season of the year i. e. if the weather is cold or warm. Perhaps evaporation is the cause.

When the sky is covered with clouds, the clouds are positively electrified. These being formed by the vapour from the (condensed) water are bodies which conduct the electricity ; which can therefore undergo and exercise certain influences on living beings.

The magnetic currents have an influence on all beings, and I think that the pigeon accustomed to fly in the neighbourhood of its loft, knows and feels this influence which is quite normal for the pigeon, and it can also discern it at any distance.

Being set free in different places it discerns this influence and benefits by it to find the direction which it must take to return to the pigeon-loft.

What makes me most certain on this point are the experiments to which I have applied the greatest attention. It required perseverance and reflection.

A fullgrown pigeon was set at liberty in a garden the entrance to the loft facing the spot with which it was familiar, it flew up 4 or 5 metres and without turning round came down on the board and entered the dovecot immediaetly, this proves that it knew the spot from where it set out.

The next day it was again set free in the same place, but with one eye closed, by means of a little piece of sticking plaster, it hesitated for a moment then set off and did not delay in finding its loft. This is even better shown in the case where pigeons come home from the races in good time, though blind of an eye, and suffering at the same time from their wound.

The day after that the two eyes were covered in such a way with paper so that it could see nothing in front — only a little at the sides, still after a little searching and hesitation it found the pigeon-loft.

Then by completely blind-folding the eyes when thrown into space it flew around and descended not far away, and came home some hours later, having pulled off what prevented it from seeing : here I noticed that the pigeon did not go too faar away from its loft.

In the following experiments things did not turn out in the same way.

The pigeon being again set free in the same place but with one ear stopped and the sight free, it flew up and turned around, and came home only three days later, when it got rid of the paper stuck to its ear.

The last experiment proves that the pigeon needs its ears to find its way home, for, the same bird being set free in the same place in fine clear weather — with its ears stopped by means of a piece of sticking plaster, only came home three weeks later, after flying up very high and turning about all the time until night-fall, which certainly obliged it to come down anywhere, it continued like this until nature helped it to get rid of what hindered it in finding its way.

So that is, dear reader, in my humble opinion what directs our pigeons in the races and during the training.

I think I may justly say that nothing can prove that it is any other sense but that of the hearing which helps the pigeon to find its way home, and it is through the influence of the magnetic currents which surround the atmosphere and the air.

However, I admit that the sense of smell assists a good deal, because, after certain experiments which I have made I conclude that some pigeons after undergoing an opration on the nose, or when through an accident part of it was broken off, became useless although they were formerly excellent racers.

Therefore the nose also plays its part and I conclude that certain fibres correspond with the brain, coming from the eye the nose and the ear which have an influence over the will-power as the bird searches the direction it must take to return to its loft.

But, that which helps the pigeon most in finding its way is the hearing.

MANAGMENT OF PIGEON-HOUSE FOR NATURAL OR PARTIAL WIDOWHOOD

Put all the birds for natural or partial widowhood in to one loft.

And for the races choose only pigeons that are sound and healthy, and which have had a good moult.

The pigeon-house should be constructed on the lines which I have indicated in the chapter on pigeon-lofts, it must always be kept dry, clean and properly aired.

It should be disinfected as well as all material in connection with it, before and after the races. Avoid draughts.

The food must consist of a mixture such as I have described in the instructions on feeding, for each branch of the hobby.

The small grain should be given in the morning, and the big grain in the evening as well as the « Combat » bread.

If there are birds in this compartment that must feed their young, it is better to increase the ration, but watch carefully to see that they do not get too fat, and that their flesh remains hard, while the birds themselves keep in good health.

(t) = They must have a bath in luke-warm water every week, the day after the race.

All the pigeons in this loft must have their daily exercise in the morning and again in the vening during one hour, the cocks first and then the hens.

During this time the loft can be properly cleaned out, fresh, clean water put into the drinking vessel, and the ration for each pigeon put into a trough inside the pigeon-hole.

Take care also that the bin with the grit contains all the necessary condiments.

As soon as the hour's flight is over, let the pigeons go in, while you stand quite near the hole, in order to tame them from the biginning, as each pigeon comes to its trough, speak to it and rub it down gently with your hand from the head to the tail.

Do this for two or three days after each exercise then as they get accustomed, they will become more familiar, and will come more asily near you, then rub the foot down gently with your fingers to where the ring is placed.

Continue this for a while then place your hand directly on the ring, squeeze it gently without, however taking the pigeon in your hand.

Accustomed to see you in the loft on their return, they will not be in the least frightened, they will all know that you are going to speak to them and to touch them.

On coming back from the competitions they will not be surprised to see you near their pigeon-holes, they will enter quickly and let you take off the ring quietly, without flying from one corner to another, like this you will gain time and money.

The pigeons for partial widowhood must be free in the loft between the hours of training in the morning and the evening.

Thus through cooing and calling the females, they will make the whole group jealous.

When they have finished their evening exercise, they must get time enough to eat and drink, and then be closed up in their pigeon-holes until the next morning, and in complete darkness if possible.

By this process, the birds will have a beneficial rest which will help when they must use all their strength for racing.

In the instructions on Pigeon-lofts for Natural or partial widowhood I have said that the holes should be 60 cm. wide ; 40 cm. high ; and 40 cm. deep, these dimensions are quite suitable for the birds set out for partial widowhood.

The fancier who knows his pigeon and has some that should have a part made to the following dimensions 40 cm. high ; 40 cm. deep ; and 40 cm. wide, the pigeon-holes should be divided into two parts by means of a board placed vertically but detachable so that when it is standing there will be two places each 20 cm. high ; by 40 cm. deep and 40 cm. wide.

Like that there is little space on top and therefore the occupants cannot have contact with each other.

Each time that the male goes after the female there not being sufficient space in the pigeon-holes the pair will be obliged to descend to the ground to mate, where it will be quite impossible unless there are no other birds in this pigeon-loft.

Every pigeon-fancier who has closely observed his group knows that the cocks are very jealous and do not tolerate any great familiarity with their « wives ».

That is the reason why those pigeon-holes should be constructed so as to have little space on top to prevent any intercourse between the cocks and hens.

Although the couple have not had any contact with each other, the female being in good health will lay two eggs as usual, and she and her partner will sit on them.

In this pigeon-hole there must be a large nest to allow the pair to sit on the eggs or even to hatch young ones if desired, so you can see which one is the best to enter for the competitions.

If you want to have young ones, these eggs not being sterile you can take eggs from another breeding loft, this method can be practised when you have a bird showing good dispositions with eggs, and then with a young one.

NATURAL

In the system of Natural you must leave the pair together, the female and the male can take part in the competitions.

The fancier whose group is distinguished, playing the
« Natural » system should seperate the sexes from the 1st
January to the 15 February continuing sometimes even until
the month of april according to the competitions in which
the birds must compete.

Those participating in short distances races should be
mated first, those, for long distances and the young bird
should only be mated in the mouth of April.

The birds for Natural are very often well disposed for
the speed competitions.

THE BEST MOMENT FOR THE MALE
IN SPEED COMPETITIONS

1) When the hen lays its first egg on the day it is
removed and that it goes cooing into another pigeon-hole.

2) When it has mated with a hen that does not lay eggs
and that you have succeeded in making it take an egg which
you previously placed under the hen, a few hours before the
departure.

3) So as to make it jealous and angry, the hen must be
put with another cock, it must even occupy the same hole,
for which there will be a quarrel, the rival is the one to
be engaged on condition that the latter comes out victorious.

4) When it is closed up in a large basket, and carried
to a quiet spot, alone for three days before removing,

without, however taking part in any daily exercise, whi'e the flesh remains hard and the body very light, a bird played in such a manner must be fed with a mixture of maze 75 p. c. ; hemp 25 p. c. (20 to 25 grammes a day) a piece of sugar about the size of a grain of maize in the evening.

If it is necessary only to remain half a day in the basket, you must give half the ration in the morning and put it into the basket if possible with the crop empty (a very important point) as it is always the most hungry ones which arrive at the highest altitude for the departure, they fly much faster and on arriving home enter the loft like « lightning ».

THE BEST MOMENT FOR THE MALE
IN LONG DISTANCES AND HALF DISTANCES

1) When it has been hatching for two or three days and that it appears to be very fond of its nest.

2) When it has been hatching for 8 or 10 days and that it stays in the nest longer than usual i. e. from 10 to 14 hours.

3) When hatching, and before putting the pigeon into the basket, you give it a young one which can eat grain on condition that your bird has no, little wings or it will have a bad moult next season, those wings must have quite disappeared before you can class the pigeon.

4) When it has a youngster of 9 or 12 days and that it keeps the nest, without looking for the hen.

5) When it has a youngster of about 20 days and that the hen is about to lay its first egg, or has just laid it.

All those conditions are excellent but be sure that the pigeon looks gay and energetic.

THE FEMALE FOR THE SPEED COMPETITIONS

To judge hens is more difficult them cocks. She can be selected :

1) When she has eggs from 8 to 10 days and she still clings to her nest.

2) When she has eggs from 16 to 17 days and she does not leave her nest only for her food.

3) When she has young ones from three to four days, and she has to provide for all their needs herself, for a day or two, she must necessarily show her love for the little ones.

Generally she does not go out of the pigeon-house any more and if you put her out she will come back again.

In this « position » you will let the male come into the pigeon-house to feed the young ones while the mother is away, when you take away the latter then let its partner come back ; continue this method as long as the hen shows herself devoted to the racing.

From the hen begins to sit on the eggs, she sees her position improving as the time approaches for the birds to come out, and then she reaches her best.

To put a young pigeon under a hen that is hatching the day before you put het into the basket, often causes her to come home quicker.

THE HEN
FOR LONG DISTANCES AND HALF DISTANCES

1) When she has been sitting on the eggs from 10 to 14 days, having taken away the cock the day before putting her into the basket.

2) When she has a young one of 10 or 12 days, until the latter is fully reared, on condition that there is no cock in the pigeon-house.

3) When, as she is hatching, you give her a little, one, which can feed on grain, the day before putting her into the basket ; taking into account the remark I made about the males and their wings.

PARTIAL WIDOWHOOD

The birds for partial widowhood can also be put into the pigeon-house for Natural.

They are mated at the same dates as those just mentioned according to the time at which they must compete in the races.

This loft will necessarily have two different sorts of pigeon-holes, as two different systems of the hobby are practised.

For partial widowhood they must be of the dimensions I have already indicated in order to facilitate the different little performances such as mating, shutting up one while waiting for the return of the other ; mating the hen with another cock with the view, to excite jealousy, etc.

During the day, when you have half an hour to spare, go to the loft quietly, and watch the birds through the little window in the entrance door. The widower will show his qualities as well as his defects. Be on your guard above all if it scratches its head, opens its bill too often to yawn, raises its feathers in front, sneezes, etc., etc.

This pigeon-house should be arranged so that it can be made dark inside as soon as the evening exercise is over ; during the races and for about half an hour after the last meal at night, so as to help the digestion and to secure a good rest after a hard days hunting after the hens ; after breeding, or the strenuous efforts made when returning from the competitions.

You must close your blind so that there are no openings to let in the light.

The repose due to the darkness for half an hour after the evening meal will help them to regain their strength. If you train your pigeons as I have described you will have them quite fit for any competition. This will enable them when racing in a North wind to cope with the group easily.

During the resting season, that is to say from the races are over until the Spring, the pigeons do not tire themselves nor make any strennous efforts, they rest from 4 O'clock in the afternoon until 8 O'clock the next morning during the competitions, they must remain several days in the basket without any rest ; must I therefore insist on your giving them extra rest. I think it useless to point out to you that the rest during the racing season is insufficient.

The extra rest which I advise you to give the pigeons will have an excellent effect, which is of the greatest importance.

If your pigeons are bred according to my instructions they will be placed under the best conditions for competing, the food distributed in the way which I have indicated will prevent them from getting too fat and the daily exercise will keep them light and at the same time keep the flesh hard.

All these conditions will enable your birds when racing in a North wind to keep at the head of the group without any difficulty : because they have been prepared or trained.

The partial-widowhood consists in taking away one of the partners either male or female after it has been hatching for 12 days on the first eggs or when they have reared a young one and laid twoo more eggs on which they have been sitting for 10 or 12 days. During this period you could have noticed, their attention to the eggs or their affection for the little ones of which the cock has sole charge, so you must use discretion.

If it must take part in a speed competition the cock must see its hen before the departure and on the home-coming.

For those selected for long distance tosses and half length ones, the hen can on no account yield before the departure.

After the home-coming the hen must be closed up in its basket on the side where its nest is placed, this must be done with pigeons engaged in speed competitions as well as distance competitions.

Once the cock returns the hen must stay there for an indefinite period, perhaps until evening, so that in this case there is not to be any fixed rule.

If the cock has endured very strenuous efforts during the race, it would be better not to let them come together but to give the cock the necessary care at once ; — a bath in lukewarm water, to wash the eyes with boracic water, etc., etc.

If on the contrary ; the pigeon has come home in good form, they can be allowed together.

That must be left to the judgment of the fancier.

Important remarks

As long as a bird played in partal-widowhood distinguishes itself honourably, let it continue the same system, if it begins to lag, stop it, and mate it again, let it lay eggs and sit on them for 12 days, the same as you did at the beginning of the season then seperate them again.

Begin the training as before, only, do not forget that this system cannot be practised in the loft for total widowhood. I shall explain the reason later in my instructions concerning this system of the hobby.

The day you put them into the basket all the birds, taking part in the speed competitions must be taken some kilometers from the loft and let out one at at time with at least an interval of 5 minutes between each.

During their absence you must place each hen in its pigeon-hole, and then remain beside the pigeon-house until the cocks return ; in order to prevent any intimacy, but at the same time to make them understand that when the come home their hens will be waiting for them.

The same method must be employed when it is the hen which is to compete.

The pigeons playing partial-widowhood need a lot of care to prepare them for the competitions.

Their state of formation is much more difficult to find out than that of those playing Natural.

The intelligent fancier who has managed the same loft for a long time, and the same stock ; will be able to see at once if the pigeon is in form.

The way the pigeons behave in the pigeon-house their posture, attitude, and pace indicate if they are in form or not.

Take one in the hand, you will know by touching it : there must be no pimples ou the rump, or under the wings and there must not be any fine down on the upper part of the crop nor on the neck.

When the bird is well disposed the muscles are strong the flesh hard it is light and not fat, the pelvis bones are quite close together and the wings are free and limber.

The droppings must be hard, close together and in small quantities, the colour is three quarters grey-black and a quarter white : if on the contrary the droppings are liquid the bird cannot be graded.

If you watch them all together, the pigeon in good form seems to be much slighter than the others the feathers stick close to each other, chiefly those over the ears, and eyelids ; the wattle must be of a pinkish colour, the feet very red and free from dandruff, and the tail very narrow.

So that the birds in the system of partial-widowhood have the best dispositions, they must be fed according to the instructions which I have given for that branch of the hobby, and all the conditions which I have just described must be carried out.

Then, during the day when you have time to spare, go quietly to the loft, and watch the pigeons through the little window in the entrance door, the pigeons which are mating will show their qualities and defects as well as the widowers.

Be on you guard about a bird that scratches its head, opens it bill, and yawns, raises the feathers on the head or forehead, sneezes, etc., etc.

When you have a pigeon which has proved its good qualities or has distinguished itself, either on its eggs or with a young one, it is not necessary to mate it a second time.

You will always succeed by working in the following manner ; —

After the evening meal when you have made the pigeon-house dark inside let the hen come into its pigeon-hole, wait till the cock is sitting nicely in the nest and calling the hen,

Then pass the eggs, or a young one underneath, on condition that the young one can already eat grain, leave the hen there for about 5 minutes.

If it does not leave the nest it is a proof that the trick is successful ; then take away the hen again.

If it has been a young one that you have put into the nest you must put some grain into the trough, if it takes some of it and then feeds the little one, you may enlist it without fear. It will retain its qualities and formation as long as it is interested in the little one.

A pigeon which has been trained in this manner must not see the hen again either before being put into the basket or after returning from the toss, only you must help it to feed the young one by giving it some soaked horsebeans occasionally.

This is an excellent method, especially for the long distance races, and for the late tosses, when you have to compete in an important competition and your pigeon must retain all its plumage : as soon as July comes if your pigeons are brooding it is more than probable they will loose their feathers in three or four days.

During the absence of the racers, so as not to frighten them on their home-coming, and to keep up the habit of entering immediately, there should be no change made in the loft.

When the racers come home during the hot weather the drinking vessel must be in its usual place, only it should be empty — about a quarter of an hour after the racers have gone in you can fill it, the water must have boiled add a few pieces of sugar or pure honey to it.

After the competitions are finished, you can leave all the pigeons together, treat them as I have described in this chapter, then be sure to take away all material including the pigeons-holes to have them disinfected and put into shelter against the rain, but so as to receive as much cold as possible.

These pigeon-holes must be replaced by little cubes (cubicals) which I have already described in my instructions concerning hygiene in the interior of the loft.

The grain must be given at regular hours and also the greenstuff, chopped, not forgetting the condiments, and also a bath in lukewarm water at least once a week.

CONTENTS

The Four Seasons

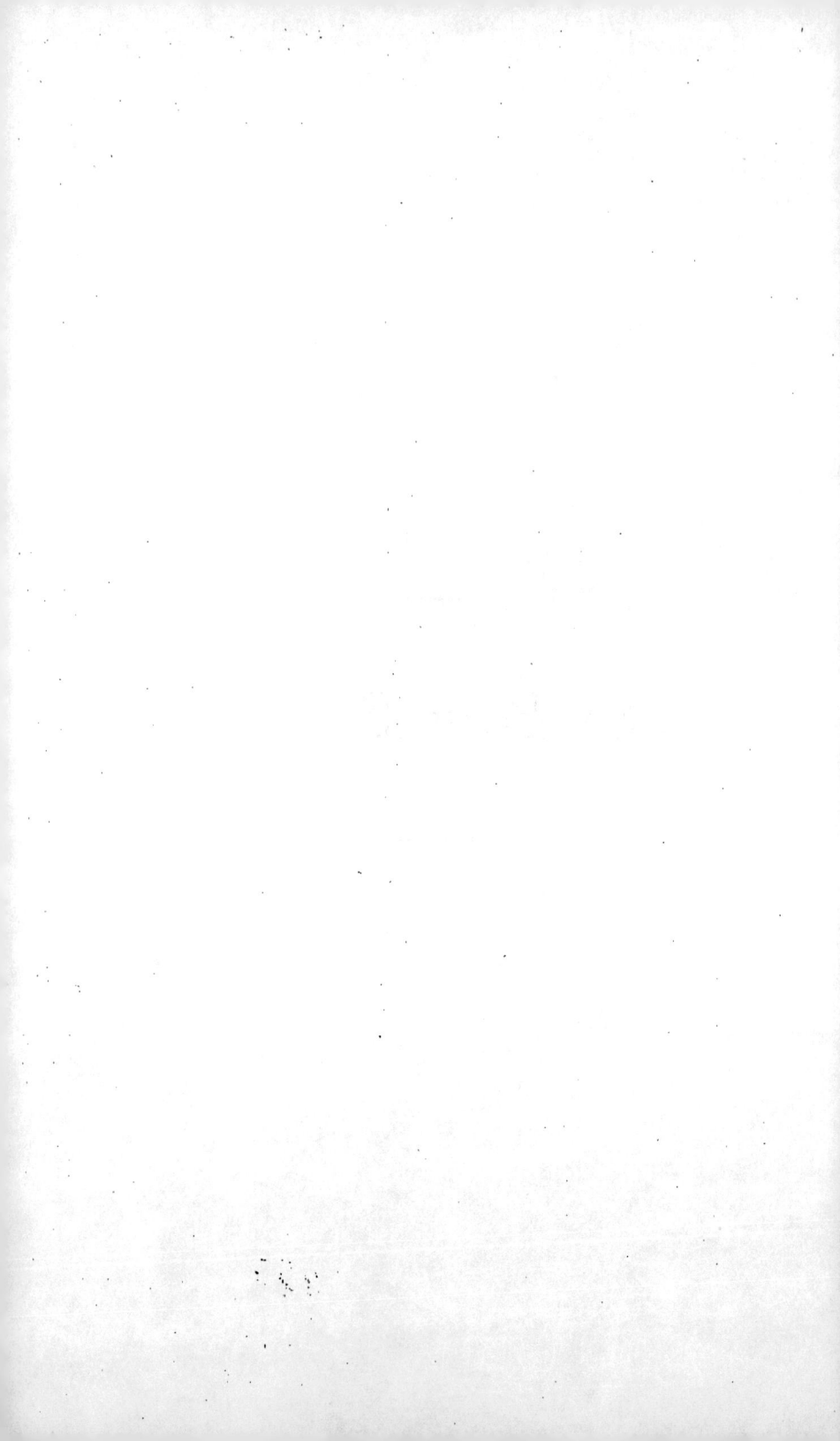

THE
FOUR SEASONS

REAL COURSE ABOUT PIGEONS

Published by

M. Joseph HEUSKIN

EXPERT IN PIGEON-BREEDING

Léopold Street, 88, FLÉMALLE-GRANDE - Belgium

Translated from French by Aug. LEMMENS

Course in four lessons, treating of the instructions about pigeon-house managing.

THIRD LESSON

The plumage and the moult — The eyes — Selection and mating — Management of lofts for " Great „ widowhood — Exercises.

Course, Number   4522

The Plumage and the Moult

The pigeon has feathers of different shapes and sizes.

The small feathers on the body, neck, and head protect it from the inclemencies of the weather.

Those in the wings and tail bear it up when flying.

Those small feathers on the head, neck and body protect the organs : but the feathers in the wings and tail form part of the flying apparatus and are consequently of very great importance, and in a degree determine the value of the carrier pigeon.

To be a good racer the pigeon must possess certain talents and their bodies must be covered in a special way, which enables them to fulfil the difficult functions which to different competitions necessitate.

During several years study and obsrevation I have made comparisons between other birds and pigeons, so as to determine the best points in the construction of their « flying apparatus » and those which give the best results.

I have concluded that the pigeon's feathers especially in the first and second wings, and those in the retrices of the tail, form, if I can say, a large book, precise and complete which it is necessary to be able to read and understand.

In this book, the remarks which I am going to point out to you, will inform you on the aptitudes of the birds, the facilities for flying, the state of health, and very often the way in which the moult is operated.

These particulars will show you if the pigeon has suffered during the race, if it has reared a little one, or if the fancier has only given it a little one while it was brooding, if it has spent one or several nights outside its loft, if it has outdone its strength ; the purity of the blood ; if it has the aptitude for speed flights or endurance flights, etc., etc.

THE FEATHERS

The feathers are composed of a stem, down and «beard» When the pigeon has had a good moult the quill in the wing can bend without breaking.

The stem is the horny part to which are attached the down and « beard » it grows out of a bulb through an afflux of blood.

It is this bulb, situated under the skin which creates the barrel or stalk of the feather; the latter is gradually transformed by the blood which contains creative substances.

It is in this barrel, or sort of enveloppe that the feather is formed by the nutritive substances of the blood.

The stalk as well as the down and « barb » form the feather which gradually comes out of the enveloppe as it receives the nutritive substance from the blood.

If you possess a good racer in your group which has broken one or several of its wing primaries or tail feathers and that it is absolutely necessary that the pigeon competes : you must introduce the new feather into this envelope, on condition that it is the same size, and forms a straight line when the wing is spread out, or open.

Instructions concerning the substitue feather will be given in the fourth lesson which treats of various confidential facts.

This barrel being bored at the base forms an air hole, and thus a tube coming from the interior of the feather, extending to the down, from this spot it takes the name of stem or stalk.

It is at this part that you can see if the birds have any hereditary defects ; split feathers (not consolidated), etc., etc. It is also in this spot that you can see if a pigeon has good or bad blood.

If the blood is bad very often this tube between the down and the part which holds the wing, is black.

A pigeon so effected cannot be graded or classed, and if you keep it for breeding the young ones will also be effected. If you have not given proper food this also can be seen on that part. So also if it has come from weak parent, it is this spot which will reproach you for the mistake. The feathers will not be united, they will be dry ; brittle, and a longitudinal gap can be perceived.

THE QUILL

The quill becomes thinner and thinner as it continues to the base of the feather.

It is this stalk, on inspection, which will prove if the pigeon has made a bad race, if it has remained a long time on the way, if you have given it a young one either to participate in a competition, or to feed before the end of the hatching term, and if, when you did so the feathers were not yet quite out of the envelope.

There will be a streak round the stem for every hard trial etc., etc., it has endured.

THE DOWN

The down which is on the feather at the spot where the stalk begins, also has its influence on the flight and on the health of the pigeon.

The more down there is, the more valuable the bird is, the closer and more consolidated the tube is, the more solid it will be.

If, when the wing is open you can see inside a very white quill covered with down, you can be sure that the pigeon possesses great racing qualities.

THE BARBES

The barb is in close rows on each side of the quill, and twines together by means of barbets, these become smaller as they approach the top.

I have already pointed out to you that the wing was a complete book, which contained much useful knowledge.

If the tube, quill, and down have their importance the barbet also plays a very important part, and affords an amount of information to the observing fancier.

When the barbet is waved it shows that the pigeon has not moulted under good conditions : if a space can be seen between, if they form a sort - of cobweb or are cut in several places the bird will be struck off the list for the whole season.

If one or several of the wing primaries have not grown to the full length the pigeon cannot be classed until those are renewed.

The pigeon which has barbets on the lower side with a tendancy to turn towards the body will be considered as incapable of accomplishing an endurance flight.

A healthy bird which has had a good moult has very straight barbets along 3/4 of the length of the wing primaries, and the others incline slightly towards the outside.

A very valuable pigeon has barbets on the long feather, that is to say on the stalk of the wing, which form a channel in the lower part, to a length of about from 7 to 8 centimeters slightly curving towards the outside.

In this wing when you see a space, or that the narrowest barbets on the outside, are, separated or divided, it is a sign that the pigeon has suffered and it will not be fit to appear on the list all the season.

DIFFERENT KINDS OF FEATHERS

The pigeon is covered with different kinds of feathers which are combined in such a way that they lap over one another so as to let only the parts of the feather which is covered with beard be seen, and these bear the name of wing primaries, second wings, coverts, and scapulars.

The ten wing primaries are planted in the follicle of the « hand », one sometimes sees birds whose wings have eleven primaries each side, others which possess ten on one side and eleven on the other side.

The secondary wings are generally about twelve on each wing and are planted in the follicle of the « fore arm » or fore wing. The coverts also grow there and cover the wings. One part, the top of the wing, constitutes the upper mantle, and that under the wing constitutes the lower mantle.

Their number and size vary in the different pigeons they cover the roots and tubes of the primaries and secondaries.

The scapulars are attached to the humerus, they cover the shoulder, they protect the tubes and roots of the wings and coverts. All these different feathers which accomplish different roles have a great effect on the health, and the value of the pigeon. The finer and more flexible the stak is, having no transversal streaks, nor lice, (according to instructions given in the first lesson on this subject, treating of the ideal conformation) the thicker the down is along the tube to the stalk, the wider and closer the barbets on the lower side are — the extremities slightly curving towards the outside, the straighter all these feathers are, contour rounded; the more valuable will be your pigeon for racing and reproduction.

THE TAIL

The ordinary tail includes twelve feathers called tail feathers. I have seen pigeon with fourteen and even sixteen ; this, as I have explained in the first lesson is not a quality but rather a defect, descended from ancient races. These feathers are planted in the follicle of the rump and enables the pigeon to controle its flight in an easy and efficacious manner.

The roots and tubes of the tail feather are recovered by coverts and by (uropygiennes) feathers, those on top cover the upper part, and are called the coat or mantle of the rump, those underneath, which also cover the roots and tube of the tail feathers and which keep out the air, are called the supporting mantle.

If I have pointed out to you the qualities and defects of the wing primaries, secondaries, tail feathers and scapulars. I must not forget also to tell you that the tail feathers can indicate qualities and defects too.

When the pigeon's tail is open and there can be seen one or more of these tail feathers of a dull grey colour, the moult has not been properly effected, and the tail has not all new feathers; also when you see transverse streaks on the quill or open spaces like a cobweb, do not engage such a bird, because it cannot be graded during the racing season as long as the new feathers have not grown.

THE BODY

The body, the head and the neck are covered with feathers much smaller than those in the tail and wings. These feathers protect the pigeon from the inclemencies of the weather.

The down, with which the whole body, except the head, is newly covered during the resting period, a so protects the body, this coating consists of a number of fine feathers and down, resembling little hairs.

THE MOULT

The moult which some prople consider as an illness, is a (natural) physiological function by which the pigeon sheds its feathers, and is again clad in a new coat.

This function is normal, natural and necessary to the proper upkeep of the organism. The manner in which it is operated reflects the state of the pigeons health. Every bird is subject to the moult.

The new feathers come on at least once a year. One even sees cases in which the new coat is not the same colour as the old one.

I have often wondered why certain fanciers say that the moult constituted a state of illness, if such were the case, how could the pigeon overcome all the difficulties ; and fatigue during the races. It happens that this natural function of moulting does not take place in exactly the same way in all cases ; some have no difficulty, while others suffer from many inconveniences, and discomforts.

It is true that when shedding the last three or four big wing-feathers i. e. when the pigeon is in the middle of the moulting process, it seems sad and depressed, on account of the different natural transformations which the system undergoes.

For this reason it is necessary to have a reserve stock of energy and strength so that the moult will be effected under good conditions.

A slight illness, an improper diet, a night spent out of doors, and many other little things have a bad effect on the moult. which sometimes stops altogether.

This is another reason for which it is better to study closely the « phenomen » in all its stages, and in every part in order to be quite at ease about the competitions for the next campaign.

DURATION OF THE MOULT

In pigeon-breedin, the essential point is that the moult manifests intself normally and continues without a stop until the end. In this way the pigeon will be always in excellent condition for future campaigns.

The moult of the pigeon lasts all the year.

From it is weaned, the moult begins and always remains.

During the first five months of the year the down is transformed into little tufts which fall off. Then the wings begin to strip, afterwards the mantle, the body and the tail.

This function, in spite of all, continues during the four seasons, but not all the time with the same intensity.

It takes place differently, and nearly always at the same periods ; these can be subdivided into the following :

1) *Winter period :* Approximately from the 1st January to the end of May during which time the pigeon loses its down and sometimes several wing primaries.

2) *Summer period :* Considered during the month of May and at the end of the voyages, during which time part of the new wing primaries come on.

3) *Autumn period :* From the tosses are over until the 1st of January during which time the last of the wing primaries come on, as well as most of the feathers on the body, the neck, the head and the tail.

It is during the Autumn period that the pigeon undergoes most of the difficulties of the moult and it is, therefore, the period when the greatest care is needed. Most of the new feathers come on in a very short space of time so it is a most critical period, and one when the pigeon requires proper attention.

This difficult period sets in during August when the competitions are over, and continues until the end of the year.

COURSE OF THE MOULT

The most critical period of the moult being immediately after the tosses most necessarily a rest, is required — holidays so to say — during which you must devote all your attention to obtain a regular moult without hitches or breaks — in a word a perfect moult.

To arrive at this result you must give the care necessary for that purpose, and I therefore advise you to scrupulously keep to the following instructions : —

1) In the pigeon-house ;

2) Manner of treating the pigeons ;

3) Diet ;

4) Sanitary arrangements.

THE PIGEON-HOUSE

As soon as the competitions are over, if you have trained your pigeons for the Natural or if they remain in the small speed lofts, the pigeon-holes must be pu'led down ; if, on the contrary you have practised partial widowhood, your pigeons must be mated again, and it is necessary to wait until they have laved the eggs and sat on them for 12 days, maximum, before that operation. For this reason I advise you, when you build your pigeon-house to use separate, detachable pieces, for the pigeon-holes.

Those parts when taken asunder should be thoroughly disinfected. When this is finished they should be placed outside the loft, sheltered from the rain and wet, but exposed to the cold.

A layer of dry sand must be shaken on the floor about 3 or 4 centimeters thick, and every day, before each meal, this must be scraped and stirred with a little rake. All the feathers and little twigs which may be thrown about should be taken away, so that there will be nothing left for the pigeons to make a nest with for themselves.

On this layer of sand you must place as many wooden cubes from 10 to 15 cm³ each, as there are pigeons, placed about 50 centimetres from each other, and from 30 to 40 centimeters from the wall.

These cubes replace your pigeon-holes, as the pigeons will come and rest on them for the night.

Above all have no flower pots upside-down on the floor such as we have seen in certain installations ; the earth is much too damp, on account of excrements, etc.

These cubes can be made from 5 little boards one attached to the other to form a little bin which must always be empty.

I advise you to keep it empty to facilitate the cleaning, and at the same time to prevent microbes from taking refuge there. For Pigeon-breeders, like myself, who have little time, being occupied with the sorting and classing of the pigeons, the best thing to do is to place these little cubes on a partition in wire netting ; holes about average size raised about 50 centimetres from the ground.

The inaccessible part underneath the partition (treillis) must be covered with plenty of quick-lime which will dry up all the droppings.

By this method no dampness will come from the wall. The cocks and hens can remain together on those little cubes. So the moult will be regular, normal, and natural.

There must not be any consoles on the walls or anything else on which the pigeons can perch.

This system is recommendable as it prevents the pigeons from struggling and making fruitless efforts which are injurious to the moult.

HOW TO TREAT PIGEONS

As soon as the competitions are over, if you have played the system of Natural, it is a good thing to separate the sexes, but it you have played the « real » widowhood and that you have seperate lofts for cocks and hens you can continue to keep the birds separated.

If you have only one loft, you can leave the pigeons together on condition to place the little cubes as is explained above, wooden cubes : and then after each meal, be sure to take away any odds and ends, or feathers with which they could build a nest.

I advise you therefore, if possible to separate the pigeons, for the fancier who has cocks, « bachelors », can see that in genaral they moult earlier and under better conditions than those that have not been separated from the hens.   ,,

However, I myself have had the experience several times and had pigeons with magnificent plumage.

The separation prevents their laying and hatching.

In such a pigeon-house it is a complete rest, a thing so necessary to the pigeons during that critical period. Do not hesitate therefore to follow my method you will obtain a wonderful result. It is so easy to set up a temporary partition, dividing the loft into two parts, it does not matter if this partition is in wire netting or closed completely, the chief point is that your pigeons retain all their strength and vigour.

So that, the renewal of the plumage be effected under good conditions, it is necessary that the body retains all the substances which may be necessary for the execution of that delicate function.

The loss of any of these substances, can cause the pigeon a shock, which may injure its health enormously and cause an irregular home-coming ; if the daily exercise lasts too long, or if the pigeon gets a fright when entering the loft, etc., the consequences can be very serious later on.

Many fanciers do not know how to controle their ambition ; as soon as a pigeon has distinguished itself, they enter it for race after race, until the poor pigeon is worn out from fatigue. Even then they submit it to breeding. However, every pigeon-breeder, with no matter how little experience, knows that the moult of the procreative pigeon stops after July, when the hatching period approaches.

It is certain that the preparation of the alimentary secretion is carried on whilst hatching and that the action is manifested at the end, or last days of the incubation.

Consequently, under such conditions, and at such a period, you must not provoke the formation of this creation, wait until the moult has been completely and normally effected.

The bird, in its wild or savage state never hatches nor breeds after the month of July. All its sufferings during the moult, will be written on the pages of the big book which the wing constitutes.

If the moult has been too slow, or has been interrupted, the wing primaries will prove it.

You can see from this that it is very dangerous to let the pigeons brood too long, when the moulting season begins — after July.

You should take care not to let your pigeons sit on the eggs more than 12 days, otherwise you expose your pigeon to a slow moult and perhaps a stoppage in the renewal of their plumage. I should advise you to take away the eggs and the nests in good time, before the incultation period commences.

Then you need not fear that the wings will droop or be weak.

DIET

If you wish to have a good moult you must give the rations which I indicated in the lesson on feeding, according to the system of the sport to which the pigeons are alloted. Do not forget plenty of green stuff which is of the highest importance at this time. Please note that the grain which I recommend you to give must not be rationed, either in the breeding loft or in the loft for racers, without taking into account the system which you play i. e. Natural ; widowhood, partial widowhood or speed competition.

When you make up your mixture for the moult, do not prepare too great a quantity at a time, but be careful to put in linseed, buckwheat and granulated (candied) sugar, according to instructions given for each branch of the hobby, not forgetting new bread on which you have poured some cod liver oil.

The grain has to be given twice a day at the hours indicated during the racing season, it should be put into little troughs having a little rim about 2 cm. wide, so as to prevent the grain from falling on the ground and getting mixed up with the sand.

During the moulting season, the ration are not to be limited, and must be given in such a way as to enable each pigeon to get sufficient food.

Look up again the instruction given in same lesson about the drinks for the moult and carry them out scrupulously if you want to obtain good results and to have no worry in the future.

During the whole resting season ; once the tosses are over until the month of march : *twice a week* the morning feed should be put into a tin box, then must be added to this, some flour of brimstone, washed (a soupspoonful to 1 kilo of the mixture) mix all well together.

There must also be a bin with a lid on it, in the pigeon-house. It has to be arranged so that the pigeons cannot go into it, but that they can easily pass their heads through. It must contain the necessary condiments : grit, powdered, bricks, ground oyster shells, old morter, etc., etc.

A little trough containing table salt should be always kept beside this bin.

« SANITATION » SANITARY ARRANGEMENTS

When the pigeon enjoys perfect health the moult will take place gradually. The new feathers are smoothe, glossy and well supplied with barb and barbets, and the quill is quite uniform.

So that the pigeons will arrive at such a state of perfection the pigeon-house should be constantly kept dry, clean and thoroughly ventilated.

To facilitate the assimilation of the nutritive substances which contribute to the formation of the blood, the pigeons at this epriod — (during the moult) need a greater quantity of oxygen than at ordinary times.

It remains therefore that the quality and quantity of pure air allowed into the pigeon-loft will have its effect on the blood.

Consequently, you should use every possible means to change the air in the loft as often as you can. But above all avoid draughts which would certainly bring destruction to your pigeons.

In giving the bath you must also be very careful ; see that the entrance door is closed beforehand, so that they will feel no sudden whiff of cold. Give the bath between 11 O'clock and noon, it is the time of day when the temperature is best.

A bath is without doubt as beneficial to the pigeon as to man.

During the moult it is not only beneficial but absolutely necessary, especially when the weather has been very warm for a length of time. In the resting season each pigeon should have a proper bath twice a week. Do not do like certain fanciers just to leave a bath in the middle of the pigeon house for the pigeons to bathe in, if, and when they take the notion.

The system which I advise you to follow is better, simpler, and surer.

Each pigeon is put into a pail, three quarters full of lukewarm water, and it is kept there for about 30 seconds.

The pigeon-house having been kept closed long enough to let all the pigeons become quite dry ; then before their feed let them take a short flight which will be very beneficial to them.

While moulting a pigeon needs plenty of liberty all the little « flights » which it takes about its loft is exercise which keeps it in good condition and health, and keeps the muscles in action.

This little bit of liberty is so necessary ; it prevents them from cramming which they will do if they are kept quite inactive. As at this time they get more food than during the racing season, it is essential that they move about a good deal, otherwise they will get fat and heavy, and the organs will become stiff.

If those organs remain inactive for any length of time they will waste away. It is we.l known that an organ does not develope if it is not made to do that work for which it was intended.

Certain movements which the pigeon repeats have a wonderful effect on the health. The birds which exercise such movements become more live'y, energetic and vigorous, and the moult is greatly benefited by them, on condition that they are not overdone of course, for over exersion can have very serious results.

I have pointed out to you above that during the resting period and the moulting season the pigeon is on holidays, therefore it is at liberty from morning till evening : what you should examine closely, all the time is the special conditions under which the moult is accomplished.

On fine days, from the beginning of September until the end of October, immediately after the morning feed, you must put all your pigeons out of doors and then close the entrance without however preventing the air from getting into the loft. None of the pigeons must enter the pigeon-house until they are called in at noon in this way the pigeons will be obliged to stay on top of the roof where they can breathe plenty of pure air, and they can also have a sun bath which is so beneficial, not to say necessary, especially at this period.

In this way you make your pigeons take much exercise around the loft, for when they are sitting on the roof the least little noise, the clack of a whip, etc., will send them flying about and thus they give movement to their organs and muscles.

The most critical moment of the moult is when the pigeon has only to shed its last big flight feather. When it loses this feather, it arrives at a stage when you should be doubly careful and prudent.

Watch your bird : — At that time it loses taste for everything, and only wants to be left alone.

It will fly only with much difficulty and will not feel disposed to take part in all the little diversions it was accustomed to.

At this time you must never take the pigeons up in your hand, as the last flight feather is very fragile and in a dangerous place, so you could easily spoil it.

Sometimes this feather still remains at three quarters its full growth in its socket. When you perceive the fact, plunge the feather into luke-warm water and scrape off the coating until you can see the little hairs. This process will make the feather open up and grow bigger.

Be careful not to force them to fly at such a dangerous moment, wait until this feather has grown to a least three-quarters its full leugth.

When the rep'acing of this feather has been effected normally, there will be no danger, and the pigeon will appear gay and eager to distinguish itself as formerly.

From the month of November and all during the resting season when the weather is misty, uncertain, foggy, rainy, or snowy, it is preferable to leave the pigeons in the loft. Here, let me add : never allow them liberty between ten in the morning and 2 o'c,ock. It is at this time that hawks are about.

At any other time of the day there is no need to dread this terrible bird of prey.

Besides these points which I have stated, there is still another to which I must call your attention i. e. disinfection. I insist that in general the pigeon-breeder does not attach sufficient importance to this point.

Disinfection ! How many fanciers ask if such a thing is really necessary seeing that they clean out the pigeon-holes and the floor every week or even every day, before the meal.

Without disinfecting, hygiene cannot exist and a proper cleaning of the pigeon-house will not be possible ; you simply take away some of the impurities but the millions of microbes which are sure to be in the holes and about the corners of the loft will remain there as long as they are not destroyed by means of a proper disinfectant.

It is certain that microbes are attracted by dirt and as you know microbes breed disease.

If you want to have your group always in good health and guarded against infections diseases, carry out carefully all the conditions which I have described in the second lesson on disinfecting the lofts regularly. In doing so you will destroy the germs which not only effect their health, but also the result of the races.

The method of disinfecting, which I have employed for over thirty years has always given marvelous results ; and it is so simple and inexpensive that I am convinced everybody who reads it will put it into practice.

Prepare a bucket of lime — rather thick — add 1 kilo (2 lbs) of table salt, a good dose of cresyline, a little turpentine, and a little soft soap, mix all together which should make about ten litres (a little over 5 gallons). You will obtain in this preparation an excellent destroyer of microbes and insects ; easily prepared, efficacious and readily employed.

As soon as the competitions are over, give a good coating, then repeat this before mating the pigeons.

If you respect those rules, all your pigeons will be in a flourishing condition, they will be preserved from infectious diseases and will give proof of their good sporting qualities, when the races begin again.

In order to point out to you how important it is to keep the loft under hygienic conditions allow me to remark that young fanciers, who do not yet know very much about their pigeons, have their birds classed in the competitions for the reason that their loft, being of recent construction, has not had time to become infected.

It stands to you therefore, dear reader to put these instructions into practice, and to reap the benefit.

HOW THE MOULT TAKES PLACE

The moult is a very important operation for all birds. It is the generator which controles the progress in all the future contests.

It has to be accomplished in proper order. Its process varies in the different pigeons and it would be vain to try to give you any definite points on this subject.

The progress of the moult depends on different circumstances, the influence of which is marked out during the evolution of that function : it depends greatly on the health of the bird ; the work which it has to perform, the diet, birth, age, temperament, etc., etc.

It is therefore most important to watch your pigeons closely and to be on your guard against bad results.

What must you do to make yourself at ease about the progress of the moult ?

Simply compare, observe, and give the food and drinks at the right times, as is described and stated in the second lesson, on how to feed the pigeons.

It is therefore prudent to make a firm resolution to watch them closely from the beginning to the end, for a good moult means a brilliant success, without interruption, for all your pigeons in their future contests.

MOULT OF FIRST FLIGHT FEATHERS

The wing primaries (first flight feathers) generally called lapwings are the first feathers which are replaced. The moult begins with the tenth feather (counting from the exterior to the interior, it is the shortest) then the shedding of the flight feathers continues in the following order.

Twenty or twenty five days after the old feathers are shed and when the new ones have grown to about two-thirds of their length, the ninth one falls off, then at shorter intervals the eighth, the seventh, the sixth, the fifth fourth, and third feathers successively fall off.

At this time, sometimes a little earlier, the coating on the wings gets undone and when the wings begin to show, the other big feathers which are on the arm and shoulder fall of.

A healthy pigeon loses the two corresponding flight feathers of each wing at an interval of one day. If the time between the shedding of these two feathers is longer than that, there is something wrong.

In your own interest seek the cause and go over the instructions carefully, which are given about diet and sanitary arrangements, so as to remedy the matter without further delay.

SECOND FLIGHT FEATHERS

The second flight feathers drop off in the opposite way, that is to say the first one on the outside and then it continues inwardly to the last one.

The shedding of the second flight feathers begins when the pigeon has only three or four first flight feathers to lose. It is in the renewal of these feathers that the variations exist.

Some birds lose only one each year, so a good number of pigeon-breeders say they can judge by these feathers the age of the pigeons, as the new feathers are generally shorter than the old ones.

This theory is false, if it is true that some pigeons only shed one feather on each wing every year, there are others that shed several, and also some others which get a whole set of new feathers on the back-wing this part can only certify that the moult has been properly effected.

SHEDDING OF THE TAIL FEATHERS

Let us see now how the moult of the tail feathers is operated. The renewal of these feathers commences when the pigeon has lost seven or eight first flight feathers, almost at the same time, or even a little later than the second flight feathers.

Generally the tail is composed of 12 feathers, or six pairs of feathers.

The fifth pair, counting from the outside towards the inside, fall first ; when the two new ones have grown to about three quarters their length, the sixth pair falls off.

And then alternatively the fifth, fourth third, first and second pairs, drop off.

The tail feathers come on when the pair just before them has grown to about three-quarters their normal length, which generally takes about ten days.

When the second pair, counting from the outside, has fully grown, the moult of the tail feathers, and of the first flight feathers is generally finished.

When visiting the houses and during the sorting of the pigeons I have often noticed that some fanciers cut off the extremity of the first pair, therefore the two outside feathers on each side would show when the moult of the tail would be finished.

However, they make a mistake, for it is the second pair which falls off last and which you must observe, to see if the tail feathers have been completely renewed.

SHEDDING OF THE MANTLE OR COVERTS

Generally a healthy pigeon sheds the coverts on the wings and the body as soon as the fifth « first flight feather » has grown.

For pigeons that must make endurance flights, the renewal of the mantel is most prejudicial.

It is the mantel which bears up the pigeon, therefore when these feathers fall off the bird is deprived of its protection against the inclemencies of the weather, and the roots and tubes of the flight feathers are not sufficiently protected.

Then the condition of the pigeon will be too low for competing or making any great efforts.

This is a very critical moment and I should strongly advise you not to let your pigeons race when they arrive at that stage.

In any case the feathers fall off in great quantities : and even, when putting it into the basket you think it can distinguish itself favourably, in decent weather, do not forget that if it has to make any great effort it will pay for the consequences later.

The coverts of the wings come on at the same time as those of the tail.

THE FEATHERS ON THE BODY

The feathers on the body begin to fall when the fifth « first flight feather » has been renewed, and at the same time as those of the mantel begin to come on.

The feathers on the body should be renewed completely and quite opened out : the new plumage, if effected under proper conditions should be thick, close, oily, and unctuous to the touch : if here and there, little spaces appear where the feather has not fully opened out, the moult has not followed its course as it should, and it will effect the next campaign.

When the small feathers on the body fall off the pigeon suffers from shock ; it becomes sad and depressed. Then it requires care, a complete rest ; no more breeding nor racing, and as much liberty as possible.

This measure of prudence is recommended through the necessity of allowing the pigeon every advantage to renew its feathers under excellent conditions, and thus guard against anaemia, and retain its health and vigour.

THE FEATHERS ON THE HEAD AND NECK

Generally the feathers on the head and neck fall off when the pigeon has shed seven « first flight feathers ». They fall off first from the head, which sometimes becomes bare in a day or two — then from the neck, the breast — etc.

The head should be completely covered with new feathers it is one of the principal parts from which the shedding of the feathers should be effected under the best conditions so that the new plumage will be abundant ; there being no down on the head, it has to be recovered with a large number of tiny feathers, so as to cover the ears, the bone in front and the cranium, in case of bad weather.

I have noticed that pigeons with a tuft on the top of the head had plenty of resistance when racing in hot weather.

THE IMPORTANCE OF A REGULAR MOULT

So that the moult be steady, it must go on gradually without a break ; it must not be forced, nor hurried, with the view to having the birds all bare together.

The moult has a most perceptible repercussion on the state of the pigeons health, at this moment, a pigeon-breeder, cannot be too careful, for here begins the preparation for the next campaign. You will never have any good out of a pigeon that has not properly moulted.

If the moult, took place under bad conditions, it is a proof that the pigeon is ill and this will be confirmed by its being inferior to its companions in the next races, and if it is submitted to reproduction the descendents of that pigeon will suffer in some way.

If the moult has been upset in any way the pigeon becomes sad, and loses many of its qualities which can only be regained when the next moult will take place under perfect conditions.

On the contrary a good moult leaves the pigeon in possession of all its qualities.

When in splendid health, it shows all its strength, and vigour ; there will be no difficulty about the coming into form, and this will last during the racing season, it will draw attention by the splendid results it will effect in the competitions. If it is destined for reproduction, its progeny will also enjoy excellent health.

To take place under good conditions, the moult must : —

Be regular : without a break and complete. So as to be regular the moult must take place unconsciously, without haste or shock, one flight feather comes on at a time, and there must be no hurry to make the pigeon lose all its feathers at the same time.

The moult must take place by degrees.

No stop : — The moult must not be stopped during its course, any interruption will be marked on the wings and the tail feathers.

The feathers will bear streaks which even with the greatest care or trouble cannot be made to disappear, they will also be very dry, slender and brittle, and there will be spaces in the barbs and barbets through which the air can pass.

This indelible mark is a sign that as long as the pigeon has not shed its wing feathers and tail feathers, the state of its health cannot permit it to come into form and thus it cannot be classed with honour in the competitions, except just by chance, or uncertain weather.

It is therefore necessary to prevent over exercion or anything which might stop the moult and cause trouble in the future.

To be complete : — When the moult has been upset in any way during its evolution, it cannot be complete, this often happens when one gives improper diet, or when the pigeon has made some great effort, or from over exersion ; draughts, a badly ventilated or unhealthy loft, or an improper site, etc.. are factors which play an important role in the completion of the moult.

An incomplete moult leaves the pigeons in a very difficult situation, when the racing season and breeding season come on.

Therefore you must use all your endeavours and earry out my instructions carefully so that you pigeons will have a good, steady and complete moult.

HOW TO CONTROLE THE MOULT

To be sure that the moult has taken place steadily. And has been completed without any interruption you should inspect the plumage in all the minutest details. The date to carry out this inspection is about the new year when pigeons born in July or earlier should have finished the moult. The first flight feathers of a pigeon must all fall off and be replaced by new ones.

The new feathers of yearlings must be more developed, lighter in colour, richer, longer, larger, and more silky than the old ones. The colour of the new feathers should always be more pronounced than that of the old ones, as with all pigeons in general, if there has not been a special cause or stoppage during the growth, the barb should be exempt from any waves or curves or tares, there must be no spaces or openings in the feathers, and the ends should be large and round, turning slightly towards the exterior of the body. There must not be any transverse streaks in the quill, nor any cracks ; it must be properly joined to the part where the down begins, and a thing that is very important, — it should be quite free from lice.

The new second flight feathers will also be of a brighter colour than the ones which have fallen off, and contrary to the first flight feathers the hairs must turn slightly towards the lower part of the wing, that is to say towards the body of the pigeon, and again the new second flight feathers must be shorter and wider than the old ones.

There must not be any transverse streaks nor splits on the quill, nor spaces between the hairs, and it should be very supple (flexible).

The tail feathers should all come on new, and be of a much brighter colour than the old ones.

It is necessary to verify, the two second tail feathers which fall off last to see that they have come on again under good conditions.

Look carefully at the quill to see if there are any transverse streaks, the joining all along the length of the quill must be thorough and there must be no spaces between the barbs or hairs.

The feathers of the mantel or covert come on all together or in parts at a time.

Those forming the upper mantel and which recover the wings must be all new, fluffy silky and glossy : those underneath the wings are sometimes only partly renewed.

I have seen many pigeons in my own group and with other fanciers, having a red or blue ink mark under the wing for years.

I will not therefore say that this is a defect, and that in such cases the moult does not take place completely for those pigeons which I had were classed to perfection.

ELEMENTS HAVING AN INFLUENCE ON THE MOULT

The moult can be stopped, delayed or quickened by different elements.

Too substantial or rich a food given at the beginning of the year hastens the shedding of the first flight feathers. Mating too early causes the pigeon very often to lose its first wing primary before march. Then when the great competitions begin, end of may or commencement of June birds too advanced in their moult have lost half their flight feathers, and are about to lose those on the body; they are out of form and cannot compete in the great competitions.

By mating, pigeons taking part in long distance races and short distances, in the beginning of march; and in the beginning of April those for long distances, you can take part in the races all the year and in any other branch of the hobby too.

Some pigeon-breeders say that the pigeon must have lost its first wing before it is enlisted, and that it is not in form until then : This is a mistake for the system of true widowhood which I have practised for several years has proved the contrary, try to keep all the feathers on the wing until the month of June, leave it in possession of all its powers and prevent it from abnormal feathering when there is an irregular home-coming.

Any disease or illness delays or stops the moult according to the degree of acuteness, and the moult will begin again only when the state of the birds health has improved or when the disease has finished its course.

When the pigeon has been hatching for 12 days, the moult stops, and begins again only when the young ones are big enough to eat grain.

When the pigeon is put to breeding after the month of July, naturally there will be a stop in the moult, but there will follow an exaggerated shedding of feathers which will have an effect when the next racing season comes on.

Too much racing and too late races interrupt the moult, forced daily exercise, when the pigeon is not in possession of all it flying powers that is to say when it is about to lose its last flight feathers — also interrupts the moult.

The moult can be hastened by putting the pigeon into a warm place where it can breathe the odour from damp hay or straw, this should be changed every day. An accident can also hasten the moult, or the fancier himself, by pulling out some of the wing feathers, or tail feathers. By this process you violate the laws of nature, and later on you will have to bear the consequences.

If you see that the moult is not taking place normally, look up the instruction, give baths, drinks and diet the pigeon as is indicated in the 2nd lesson (food, drink). This method helps nature : therefore it is the best way to have a successful moult.

THE YOUNG ONES MOULT

The moult of the young pigeons does not take place in the same way as that of the old birds.

I have already said that this function does not terminate according to any fixed rules but that it is controled by different things, age, hygiene, food, and proper care.

For the thorough completion of the moult, the young pigeon is not only effected by the conditions which I have stated, but the date of its birth has its influence also.

Pigeons born in Januari, or in the beginning of February, at the age of two or three months, will first have a partial moult of the neck and head feathers; the moult then stops during all the racing season, but will continue its course in the late season, and finish by the end of the year.

For this reason I advise you not to mate your pigeons too early. A moult which takes place twice, or has long intervals between must be considered as abnormal.

Generally, pigeons born between the month of march and the end of June, or even in the beginning of July, are always well fitted for a normal and complete moult.

Does not a bird in its wild state (therefore its natural state) choose this season to build its nest and rear its young ? It is certain that nature arranged things much better than man can.

The birds thus in its wild state « knows » the time to renew its plumage much better than we do.

Then, why breed young ones too early or too late ? Why transgress the laws of nature ? Pigeons born between the end of July and the beginning of september, will only have a partial moult.

Certainly they will retain the feathers in the tail and wings, and these will only be renewed the next year. A young pigeon that has kept one two or even three flight feathers will have a complete and normal moult the next year.

Although it has not renewed all the feathers in the tail and wings in its first year, that will be of no importance during the competitions. Therefore when you wish to enter a pigeon for the races, which has only one, two or three feathers on each wing, you can easily do so ; it never happens that the bird loses two feathers at a time and in different places.

But if the pigeon has retained four of its « nest » feathers, or more, watch it closely the next year when it sheds the first flight feather, it will also shed the feather which should have fallen off the year before. From the time the moult was interrupted, and as it gradually continues, it will lose two second flight feathers from each wing and so on till the tenth wing feather has been renewed, then the moult will continue one feather at a time from each wing until it comes to the wing feather where the moult stopped the first year.

Therefore a pigeon that has not had renewed its first flight feathers, the year it was born will lose the short wing feathers twice, whilst those which it had retained the first year will fall off only once the second year.

Such a bird must therefore be two years old to have a complete and normal moult with no spaces or openings in the wings, when it will have lost one first flight feather.

The same conditions concern the tail, if the pigeon has not shed all its feathers the year it was born.

Pigeons born in the late season, generally lose no wing feathers nor tail feathers, the moult begins the following year, at the same time as that of the old birds, and takes place normally and regularly.

Late bred pigeons must not do fatiguing work the first year ; and not being yet fully developed, they should not be mated ; they ought to be trained only to fly short distances.

When they are two years old, and the moult has taken place under good conditions, the pigeons can participate in the tosses ; you can even let them fly distances of 500 kilometres. If they have the necessary qualifications for reproduction, you can let them do so, as well as those born at the proper season from March until the beginning of July.

SPLIT OR CRACKED FEATHERS

Any feathers that are split or cracked, or that have little grooves, are bad. They show that the pigeon has suffered, for want of proper food at the time of their growth. They denote a weak constitution, probably inherited from the parents, or an insufficient or incomplete diet, a fatiguing race when the old feathers were falling off : or their being pulled off ; and also on account of wounds or sores. If the feather has been struck at the base, it may be bent, split or broken, etc.

In this case the feather will improve during the moulting periods which follow.

If the pigeon has several feathers effected, if you follow my advice about what to do during the moult, you will perceive an improvement after each moult.

When you have a pigeon with several wing feather afflicted — for instance if there is a large opening on the outside of the quill you should not submit such a bird to reproduction, especially if it be a hen, as the hen plays 75 p. c. in the role of reproduction : keep only perfect pigeons. The sport of pigeon breeding has made too much progress at the present time, and it would be ridiculous to have any but the best pigeons in your pigeon-house.

The Eyes

The eyes are the mirror of the soul ; they are of the greatest importance in pigeon-breeding, it is through these organs that the fancier can see and learn everything.

It is these organs which guide the faculties for finding its whereabouts, they tax the instinct, vigour, endurance, and the reproduction.

They reflect, bad health or good health ; defects as well as qualities.

Quite a number of pigeon fanciers tried to explain the worth of the eyes, but I think few of them could bear up their statements ; works about that subject have come, from many pens, as have also articles about the « homecoming instinct. Through application you may have proved the veracity of their theme, but I, myself, have the pleasure to be able to tell you that by carefully following my instructions, you will obtain long desired results from those pigeons mated with the view of producing good eyes, and also from the unions constituted for the purpose of creating good racers.

I want to teach you how to make your pigeons triumph at all times, and in all winds.

Owing to long experience, I am able, on examining a pigeon, to say in which pigeon-hole it was born, if it is a she or not, and more than that, in what sort of weather it will fly fastest — a secret which nobody else knows up to the present.

The eyes enlighten me on that point, and many fanciers have put me to the test during the sorting and mating meetings, and how astonished they always were, at my answers to their questions — I was never mistaken.

Besides the physical characteristics, considering them as perfect, if the eyes do not also contain certain qualities, it is more than probable that the pigeon is not a sire, and it will never be the right type for racing.

It is according to this principle that the first sortings are carried on.

You should attach the greatest importance to the examination of the eyes. There is a song which says « The eyes are the windows through which you can see the depth of human beings ». I think this can be applied also to pigeons, for they denote such important points, that nearly every pigeon fancier, really anxious about the management of his group, and wanting to obtain something special in reproduction, should know every thing about them.

I have told you in the first lesson, on ideal conformation that only pigeons representing the real type should be united.

The same is to be said when mating for reproduction which should only be between pigeons having perfect eyes. So as to enligthen you on this point and to let you know the easiest method of treating pigeons suffering from eye diseases (see fourth lesson).

I am now going to describe the formation of the eye.

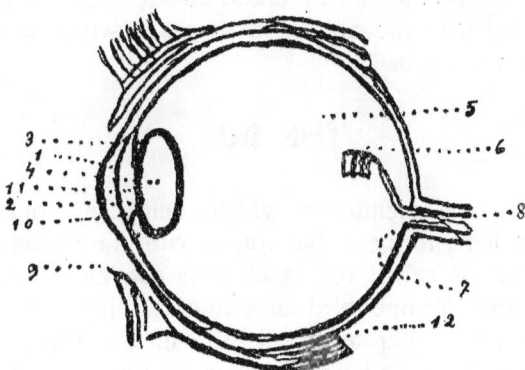

1. Sclerotic.
2. Aqueous humour.
3. Iris.
4. Crystalline lens.
5. Vitrous humour.
6. Comb.

7. Retina.
8. The optic nerve.
9. Eyelid.
10. The transparent cornea.
11. The pupil.
12. Choroid.

THE SCLEROTIC

The sclerotic is white opaque and very strong, that part especially is the most developed of the fibrous membrane, a cartilaginous lamina is formed in its depth and becomes ossified in the circumference of the cornea and the optic nerve.

These ossified parts, forming sorts of little scales are mobile and permit the orb of the eye to move about and be protected at the same time.

AQUEOUS HUMOUR

The aqueous humour is a transparent liquid, colourless, and occupies the space between the frontal part of the crystalline lens and the cornea. The orb of the eye is in the socket which is placed in the skull, and this is protected in front by the eyelids, and by the sclerotic, supplied, as I have already said with movable ossified parts, which enable the orb to change form.

THE IRIS

The iris is a membrane which, being in front of the crystalline lens, forms a diaphragm, with an opening in the centre which is called the pupil. It is convex in front; the colour varies in one bird and another, it is determined by two kinds of pigments one of which is placed in the anterior socket or bed, and belongs to the class of « lipochromes ».

Amongst which is found the « tuteine » which makes the yellow colouring of the iris; the other is placed in the deeper sockets and is formed through the « melanine », and resists the action of certain agents, while the former is extremely changeable, and the varieties of colours which can be detected in the eyes of different birds are the result of an eye disease inherited from an ancestor. The iris has to play a great part in the function of dilation and contraction, thus regulating the quantity of light following the net transmission of objects on the retina. When you have a pigeon with a weakness of the iris, the result of being pecked in the eye while in the basket or an accident, you will not be able to class it until the iris has been completely healed and regained its bright colour again.

THE CRYSTALLINE LENS

The crystalline lens is the form of a transparent lentil, is situated just behind the pupil, between the vitreous substance and the aqueous humour.

It becomes flattened out, or bulges according to the distance of the object the image of which has been transmitted to it. It is the crystalline lens which concentrates the rays of light and allows the image of the object to be stamped on the retina, it is more developed when the pigeon flies fast.

THE VITREOUS HUMOUR

The vitreous humour is covered by a white thick membrane, it fills the greater part of the interior of the eye, from the retina to the frontal part of the crystalline len, it is composer of a gelatinous liquid, ressembling melted glass.

In the front part of the orb of the eye, there is a circular opening in which the transparent cornea is enclosed.

THE COMB

The comb is a glandular organ, and is made up of a plexus of muscular fibres. It controles the ocular tension and preserves the retina from the extra strong light; it begins in the choroid just where the optic nerve penetrates, it has a triangular folding blade with pigments which sinks into the vitreous body and continues in the direction of the crystallin lens, but does not reach quite so for. It helps the sight wonderfully, and has the special property of filling up with blood and thus producing different pressures which propitiate the interior of the eye.

THE RETINA

The retina is composed of a number of layers of stratified elements. Its exact position corresponds with that of the expansion of the optic nerve which bears the name of papilla and which forms a very dark point.

It is divided into two parts, each of which is prolonged from the expansion of the optic nerve to the circumference of the crystallin lens. The anterior is devoid of visual aptitude, and the back part, seperate from the former is the so called retina.

OPTIC NERVE

The optic nerve has the shape of a piece of cord, whitish, in colour it serves as a conductor to the feelings and as a faculty for testing the moral impressions or feelings. The pigeon has this nerve divided into two parts by a ligament called the « tractus », which enables it to detach itself from the optic obule, and which through its different movements intersects the two nerves; these also separate into several little leaves which finish off at different lengths.

The optic nerve is fibrous; it is attached to the sclerotic and continues, becoming gradually thinner until it passes underneath the comb and perforates the choroid. which, through a ligment forms a sort of blade which becomes intertwined with the fibres on the retina.

EYE-LIDS

The pigeon's eye is covered by three different lids.

The two outside ones are visible, the third is situated under the upper one.

The upper lid is much smaller than the lower one, and not so mobile, so that when the pigeon closes the eye, it is the lower lid which meets the upper one and covers nearly all the cornea, it has a much larger muscle than the upper lid and is attached to the bottom of the cavity, of the orbit.

The third lid is situated under the upper lid, it is very wide and easily covers the whole front part of the eye, it is called the « nictante » membrane, it has two special muscles behind the bulb which work by means of branches, and take a very peculiar course. It consists of a very fine membrane almost transparent, its work is to continually clean out, the ocular globe, and to cover all the front of the eye when that organ is unemployed.

The ouside eyelids are croated or lined by a transparent mucuous membrane.

It also covers the orb or ball of the eye, and is devoid of glands.

THE TRANSPARENT CORNEA

The vitrous or transparent cornea is a membrane which allows the rays of light to enter, at the same time it closes up the hole in the front part of the sclerotic and completes the envelope of the ocular orb.

It is convex and is set in the front part of the orbit of the eye; its refractive indication is about the same as that of the aqueouse humour.

THE PUPIL

The pupil, being apparently round, contracts under the influence of light, it dilates or contracts according to the intensity of the rays of light, and the distance at which he objects are situated upon which the sight is fixed.

The fibres are streaked.

The pupil has a black border or edge or sometimes when the eye is perfect it is blue, it is like a quarter of the moon, which makes some fanciers suppose there are some uneven parts in the pupil but in reality circle of adaptation is incomplete.

A strong light causes it to contract while a weak light makes it dilate.

THE CHOROID

The choroid is the back part of the eye, all dark without spots, it has a plexus of smooth muscular fibres, in the lower part. The anterior is much thicker than the upper part.

The choroid participates in the composition of the « trabeculaire » tissue formed by the fibres from the iris and the ciliary muscle.

The visible portion of the pigeon's eye is formed of a number of circles which are as follow : counting from the centre :

1. The pupil.
2. The circle of adaptation.
3. The circle of correlation.
4. The iris.
5. The fifth circle.

THE EYE AND ITS DIFFERENT CIRCLES

THE PUPIL.

The pupil is an opening in the centre of the eye and through it the rays of light pass. Through the pupil we can learn many things when the pigeon is coming into form (see fourth lesson).

THE CIRCLE OF ADAPTATION.

The little black thin or bluish circle which encircles the pupil is called the circle of adaptation, many pigeon-breeders mistake it for the circle of correlation, « adaptation » means the modification, of an organ or making it more suitable for its work.

The finer this circle is, the better it facilitates the contraction and dilation of the pupil, according to the intensity of the rays to which the eye is submitted.

It makes the pigeon more apt in flying and more valuable for reproduction.

THE CIRCLE OF CORRELATION

The circle of correlation signifies the relation of two things one of which logically calls up the other. This appellation applies itself to the definition.

To this effect, you will notice that the iris contracts or dilates so as to modify the pupil.

If the iris dilates, necessarily the pupil contracts and seems smaller. If, on the contrary, it contracts, the pupil becomes larger : between the pupil and the iris, there is a relation ; constant and inevitable.

THE IRIS.

The iris is the fourth circle of the eye it is formed by the large outside circle at a certain zone, it is the latter which indicates the different colours.

FIFTH CIRCLE.

In general every pigeon has the fifth circle. It is very thin, can be seen on the outer part, all round the iris, and it is generally the same colour as the circle of correlation, or sometimes it is of a very dark colour.

THE VALUE OF THESE CIRCLES.

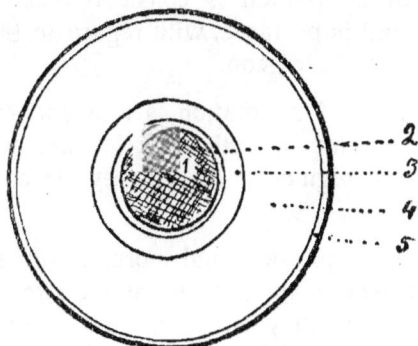

1. The pupil.   3. The circle of correlation.
2. The circle of adaptation.   4. The circle of the iris.
5. Fifth circle.

THE PUPIL.

The pupil should be very round and preferably of average size, this is better, for the carrier-pigeon subnutted to the distance competitions.

When it is large, it is a sign hat the pigeon is suitable for short tosses.

When it is smaller it shows that the pigeon can distinguish itself in long distances and half-lengths.

When the pupil is deformed, that is to say, oval or oblong the pigeon is neither suitable for racing nor for reproduction. It must be done away with.

Do not bowever misunderstand this point for instance do not think that the pupil is ugly or misshapen when it is only the circle of adaptation that is not entire (a quarter of the moon) in this case the pigeon can be quite successful in the races.

THE CIRCLE OF ADAPTATION.

The circle of adaptation is considered as beingof the greatest value and importance, with regard to the appreciation of the eye of the pigeon.

When the circle of adaptation is thin and encircles the pupil completely and the other circles are well marked and exempt from pigments, etc., the eye is considered as perfect.

If possessing the physical and moral qualities as described in the first lesson, the pigeon has the circle of adaptation thin and completely encircling the pupil, it will be considered as a type of high origine of view suitable for racing and reproduction.

Of two pigeons having the same physical qualities, the better one for reproduction and for racing, is the pigeon that possesses a circle, complete, thin, and encircling the whole pupil.

Therefore always be very careful. When you want to introduce a new pigeon into your group with the view to reproduction.

Choose the bird with the cicle of adaptation well marked, encircling the whole pupil, and as fine as posible.

A pigeon which has not this circle at all will be classed as unsuitable for the competition or for reproduction, just the same as a bird with a deformed pupil (oval).

You must be very careful when looking for the circle of adaptation, because it is often of the same colour as the circle of correlation, or as the pupil, which makes it very difficult to distinguish, especially for fanciers with little experience.

The easiest means of discerning it is as follows.

When you hold the pigeon facing the light of the sun, the pupil contracts and allows the circle to be seen, you can also hold it away from you and draw it quickly towards you several times in succession, so that the pupil becomes dilated and therefore renders the circle more visible.

Generally, when you do not see it at first sight, and if the circles are not well defined, and the exterior zone is pigmented or spotted you may be sure the circle of adaptation does not exist and that you must do away with the pigeon.

Some pigeons have only a third of the circumference of this circle, others a half or a quarter; a perfect type of pigeon has the circle complete.

Pigeons possessing only part of this circle, if they are formed under good conditions can prove excellent racers, but cannot be considered as good sires, not possessing the qualities necessary in the eye.

THE CIRCLE OF CORRELATION.

The circle of correlation also is of great importance, when classing pigeons destined for reproduction of for racing.

However there are some pigeons which do not possess this circle. Birds of this kind have the iris nearly the same colour as its « decorations ». An eye with a defective circle of correlation is called a « full » eye.

Generally these are the choice sires, which are easily mated with the view to producing high class pigeons, both for racing and reproduction, on condition that they have the circle of adaptation well marked and limited.

The colour of the circle of correlation is generally orange or white, in the perfect type it is of average size, the birds for long distances and half lengths has it narrower, and it is more marked in pigeons which can be classed for speed competitions.

When the circle of correlation spreads over the exterior zone (the iris) it denotes that the pigeon is unsuitable for racing and reproduction.

THE IRIS.

The iris is the most developed zone and constitutes the fourth circle, all the colours of which it is constituted are good on condition that they are uniform, and free from pigments and spots, etc., etc.

A deep coloured iris has a favourable influence on the pigeon ; it denotes something extra in reproduction, and renders the pigeon apt to distinguish itself in all distances.

A light colouring shows that the blood is rather weak.

A pigeon with a clear delicate iris can be an excellent racer, but the pure blood required for reproduction is lacking.

There exists in the eye of our pigeons different shades of colouring, from grey to dark red, pale yellow or orange brown and blue etc. The last mentioned is rare. but pigeons possessing this colour can be classed in the first rank of birds to be mated for procreation.

FIFTH CIRCLE.

The fifth circle is situated at the extremity of the orbit of the eye, and encircles the iris completly. All pigeons have this circle, the middling class as well as the very valuable types, only often it is difficult to see it until the pigeon has come into form.

THE EYE OF A VALUABLE TYPE.

I have already pointed out to you that the eye is the mirror of the soul. In general we can say that the eye pictures or reflects the worth of the pigeon itself.

Taking into account the bodily conformation and the plumage, it is also through the eyes that you can distinguish a good sire.

It is the eyes which facilitate the selection of the birds for mating in order to have choice young birds. It is the eyes which enable you to make unions so as to be able to have first class racers in all weathers.

It is through the eyes that you can determine its home coming faculties, reproduction, and in a word its real worth.

However before pronouncing a definite opinion on the worth of a pigeon, you must take into account its physical and moral constitution, which should naturally be first rate (see first lesson). A pigeon with eyes possessing all the qualities indicated in the lesson on « conformation », will be a valuable bird.

During my many years experience and observation. I have always seen that a very close relation existed between a good conformation of the different organs.

The moral constitution depends on the physical constitution and vice-versa — as « Vieux Bleu » has so well explained in his articles.

THE POSITION OF THE EYE.

A pigeon of high origin has the eye well set in its orb, but not sunken, only projecting somewhat.

1°) The fictitious line, extending to the slit between the two mandibles of the bill.

2°) The depth between the skull and the upper part of the eye.

To be perfect, the eyes of the pigeon should be placed so as to see in all directions. Inclining slightly towards the front, with a tendency to look downwards.

They should be as high as posible ; that is to say near the top of the skull.

When the eye seems to be divided by the fictitious line, extending between the two manibles of the bill, it is placed too low down. The pigeon can be a good racer but its qualities as a sire are not many.

If the eye is too far back, or too much towards the front, or too low down, the pigeon is unsuitable for racing and for reproduction.

Figure 25 shows a head on which I have indicated where the eye should be placed, line n° I. The space between the upper edge of the eyelids and the top of the skull n° 2 must be slightly flattened, and has to be very small. This is the reason why there are so few pigeons which give satisfaction in the races, or win prizes or even obtain « satisfaction » in exhibition competitions.

The smaller this space is the better the pigeon's intellectual and moral qualities are.

The nearer the eye is to the top of the skull, the better it will prove its will-power, home-coming instinct, memory, etc., etc.

DIFFERENT VARIETIES OF EYES

First of all, I must tell you that all the colours are good, as long as they are clear and the colour of the iris rich, and deep.

Experience will teach you that the more complete, and defined the circle of adaptation is, the more renowned will the carrier-pigeon's qualities be.

Then again, the more perfect the circles are, and the thinner and more defined the circle of correlation is the greater value the pigeon will have. If the colour of the iris is rich, no matter what the shade, it is a good sign in a racing pigeon.

An ideal type of pigeon, must have the circle of adaptation whole, the circle of correlation fine and its boundary limited and the colour of the iris must be rich no matter what its shade.

The colouring itself is of no account. There are good and bad pigeons to be found, with all the different coloured eyes possible.

Let the eyes be red, garnet-red, orange, brown, silver-coloured or blue, it has no effect on the qualities of the pigeon.

What is most important is the form of the circles, and if they are well marked and limited or not.

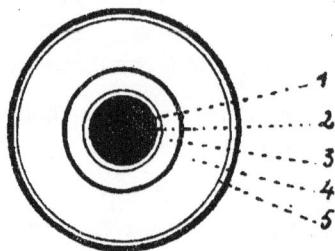

The figure nᵘ 26 represents the perfect eye, that is to say that of an extra good racer and sire.

The essential qualities which it possessess are that each circle can be distinguished separately, they are all well marked, and complete, in pigeons of this category.

The pupil is quite round and of average size, the circle of adaptation very thin, encircling the pupil completely, the circle of correlation is clearly marked and defined, the iris of a bright rich colour, the fifth circle is quite visible, which proves that the bird enjoys perfect health.

It is well to remark that such a type of pigeon has qualifications which bring success in races of any distance, when the pupil is of average size it is a sign that the pigeon can compete, in all the distances. When it is smaller it shows aptness for tosses of great length, and half lengths, when it is larger the pigeon can genarally be classed in races of short distances.

Figure 27 represents the « full » eye, which indicates a splendid sire, such as in the preceeding figure.

Besides if the pigeon possesses good physical qualities, and has silky and abundant plumage, it will also be considered as a great racer, This sort of eye can exist in all colours.

The pupil is round and of average size the circle of adaptation thin, it encircles the pupil completely, the iris is of a bright rich shade occupying the whole part, from the circle of adaptation to the outer circle of the orbit of the eye. The fifth circle is quite visible, which proves that the pigeon enjoys perfect health.

As is the case when the eye has fine circles the pigeon can possess qualifications for races of different lengths according to the size of the pupil.

Pigeons with this type of eye being generally classed as first rate sires, can also distinguih themselves honourably in long distance races.

The eye not having the circle of adaptation whole.

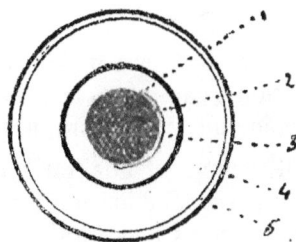

Generally pigeons having the eye indicated in figure n° 28 (an eye with the circle of adaptation incomplete) are great racers, and can be easily classed in the competitions, but they have not the qualities required for reproduction : the circle of adaptation is incomplete.

The pupil is round, average size, and the pigeon is qualified for long and short races, according to the dimensions of the pupil.

The circle of adaptation does not entirely encircle the pupil, it covers only a third, a half, or even a quarter of the circumference.

These different types of eyes denote the proportional value as a racer or as a reproducer when you have to mate it with another type, possessing the necessary qualities, that is to say one with a full eye, or having the five circles.

The circle of correlation is clearly marked and defined. The iris of a rich shade, unspotted, and free from pigments.

The fifth circle is quite visible, which shows that the pigeon has great racing qualities.

An eye completely devoid of the circle of adaptation

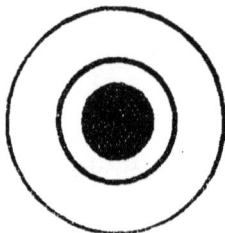

Pigeons with this eye — figure n° 29 (an eye devoid of the circle of adaptation) can race, and be graded in the competitions, chiefly in fine or favourable weather.

Being devoid of the circle of adaptation, they cannot be mated with the view to reproduction but they can be entered for the racing competitions as long as they give satisfaction.

The pupil is round and the size is in proportion to the distance for which the pigeon is capable of competing, the circle of correlation is well marked, and defined encircling the pupil completely without spreading over the circle of the iris.

The iris is uniform, of a rich colour, bright unspotted, free from pigments and well set into its orbit.

The fifth circle is quite visible when the pigeon is in form.

An eye having pigments, and devoid of the circle of adaptation

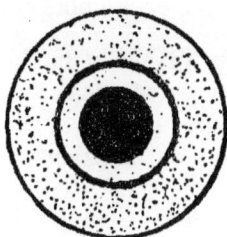

Pigeons with eyes such as is shown in figure n° 30 (an eye with pigments and no circle of adaptation) should be considered as unsuitable for the competitions, and cannot be kept for reproduction.

Such pigeons have lost the qualifications required for good racers and good sires.

They are totally devoid of the circle of adaptation or there may be only just a part of it remaining, the circle of correlation overlaps that of the iris.

Although the pupil is quite round and of the average size, the circle of adaptation being defective, the circle of correlation, and of the iris also, being all spotted over, this thoroughly depreciates the pigeon.

The result will be the same when the eyes are pale, or bloodless, when the circles overlap one another, if the eyes are badly placed too far back, too much to the front or too low, or any blemish which can effect the sight.

An eye with the pupil deformed (oval)

Nmber 31 shows an eye with a deformed pupil, oval shaped, although the circle of correlation is clearly marked and the circle of the iris of a good colour, light and rich.

A pigeon with such an eye must be rejected though it be a descendant of a type of great value or of the highest origin, it fixes only one eye at a time it does not see clearly and will have to be banished from the pigeon loft.

I again call to mind the fact that great difficulties sometimes arise when you must distinguish whether the pupil is oval or just elongated, because the pigeon may have the circle of adaptation only half or quarter the circumference.

In eyes streaked in differend colours, or dark eyes the circles can be visible or invisible.

When one of them presents the circles, you can certainly judge the pigeon after the indications which I have already given.

If the eyes are quite black, you must judge according to the physical conformation, and take into account the plumage which will show all the qualities as well as the defects of the pigeons.

Such a pigeon can belong to any of the classes cited above, from the perfect bird down to the unsuitable one. Only this colour being a sign of atavism shows that the necessary qualifications of a sire are missing, therefore I advise you to make a racer of that pigeon.

THE CONTROLE OF THE PIGEON'S SIGHT

A pigeon can possess all the intellectual, physical, and moral qualities, have well formed eyes; be an excellent racer, and still not be capable of regaining its loft when carried a few kilometres from the pigeon house.

Although you cannot see any blemish on the eye, it may still have some defect.

Very interesting results have been effected in ocular researches but nobody has yet been able to express an opinion only from the exterior of the orbit of the eye — the only part which when holding the pigeon to the light can show whether the sight is good or bad.

Many pigeon-lofts are supplied with blue glass, so that you can see what is going on inside without being perceived by the pigeons and as the blue rays are dominant in the evening when you have a lamp the pigeons sight is totally dimmed, you should put on an opaque curtain, a slide, or shutter opposite, so as to shade the light and rays which prevent you from seeing everything that is happening inside.

It is a well know fact that blue effects the pigeon's sight and, dims it completely. This was proved by throwing some grains of maize on a red, yellow and blue surface.

When it was on a red or yellow ground, all the grains were picked up in a few seconds.

Whereas from the blue ground, the pigeons only picked up the grains in spots where it was very light. Therefore when inspecting your pigeons in artificial light, only the blue colour can hurt the pigeon's sight ; there exist certain processes which generally give satisfaction in finding out if the pigeon's sight is effected or not go into the loft at night with a light — not too strong — observe each bird when it flies to the bottom of the pigeon-house to go into the pigeon hole the pigeon holes must only remain in the loft during the racing season) if they miss the board, or often mistake their pigeon holes, you can be sure that the sight is bad.

Then during the resting season, the period when the pigeon holes have been replaced by little cubes, at distances of 30 or 40 centimetres from each other — if the pigeon stumbles over the little cubes or anything else in the way, or if it picks up only the large grains, chiefly the maize, it is certain that such a pigeon is not to be kept in the loft, as it has not the qualifications for reproduction.

In order to be quite certain however, repeat those visits several times with lights of different intensities, sometimes very brilliant and sometimes weak.

Put a red paper or a yellow one on the glass, to produce different colours, it is one of the principal and most practical ways of finding out if the sight is good or bad.

When the eye secretes too much liquid from the vitreous humour, or when the eve is shaded, it is a sign that there is something the matter with the sight.

Very often a pigeon with bad sight fixes one eye at a time, while leaning the head to one side. It will not take flight when the sun darts forth its rays, during the daily exercise, it will remain on the top of a roof because the light dazzles it and it finds it impossible to, follow its companions during the exercise.

Its favourite grain will be maize because it is more easily seen than any other grain.

These few hints which I have revealed to you are of the greatest importance.

However I must point out to you that in case of illness, naturally the sight becomes weaker. To be sure of ascertaining if the sight is really good, you should observe only pigeons in the best of health, a pigeon not having good sight, or affected by any sort of disease should be expelled at once from the loft, no matter about its great physical qualities.

An exception can de made when the imperfection is the result of an accident, being pecked in the eye etc., which is certain to get cured, even if the eye has been knocked out the pigeon possess the necessary qualities for reproduction.

CLASSING OF PIGEONS ACCORDING TO THE EYES

Without taking into account the position of the eye and the colour, the classification which I am going to describe cannot be quite exact, for the ideal conformation which I pointed out to you in the first lesson on the qualities necessary for a first rate carrier pigeon. I shall draw your intention simply to the qualities of the eye and how the pigeon is classed according to its qualifications.

This classification is intended to teach you how to inspect pigeons exhibited in baskets without taking them in your hands, in order not to disturb them all when you must « break in » a new pigeon to the loft.

Classification according to the eyes can be subdivided into seven different categories :

1) The perfect carrier-pigeon, that is to say a first rate racer and a first rate sire.

2) Sire only.

3) A pigeon possessing qualities for short distance competitions.

4) Pigeons distinguishing themselves in long distances and half-distances.

5) Pigeons which can be classed with advantage in speed competitions.

6) A pigeon of high origin, wounded in an eye or (physically) deformed, or defective plumage.

7) Irregular pigeon, that is to say a pigeon which is rarely classed, and which comes into form with difficulty.

A PIGEON, PERFECT IN ALL RESPECTS.

A pigeon with the five circles of the eye visible and well defined, the pupil quite round, having the circles adaptation and correlation as thin and entire as possible, will be classed as first-class. The iris is also rich in colour, and the fifth circle completely encircles the outside zone of the orbit of the eye.

Amongst those pigeons classed as first rate in all respects and which have all the necessary qualifications for reproduction there are some which can be classed with satisfaction in all the races, whilst there are others which can only be distinguished in « full-lengths » and « half-lengths », and others again, only in speed competition.

If the pupil is of average size, the pigeon can take part in races of all distances if it is smaller the pigeon will be considered as being capable of distinguishing itself in tosses of full length and half lengths; if it is larger it can be classed with honour in the speed competitions.

FIRST RATE SIRE ONLY.

When the eye possesses the qualities given above, and it cannot compete in the races on account of an accident to the eye, or if some bodily injury prevents it from taking part in rapid flights, it can be classed as a first rate sire only; the pigeon with full eye (figure 27) which is devoid of the circle of correlation is classed in the first rank of sires.

Other indications will be given in the fourth lesson treating on the classification of sires according to my method.

PIGEONS FOR ALL DISTANCES

The pigeon with the average sized pupil, quite round and the circle of adaptation visible but not complete is of this class.

The circle of correlation is well marked, and limited, the iris is of average size, of a bright rich colour, without spots, free from pigments.

The fifth circle is very thin and quite visible.

PIGEONS FOR FULL LENGTHS HALF LENGTHS

Here can be classed the pigeon with a very small pupil, but quite round, the circle of adaptation well marked and with as much of it visible as possible in front and on the top, but not complete. The circle of correlation is clearly marked, and defined, and very thin the iris is large and of a bright rich colour, free from pigments, the fifth circle is scarsely apparent, but the trace can be seen if you raise the eyelid.

PIGEONS FOR SPEED COMPETITIONS

Into this category you can place pigeons with very large pupils, quite round, the circle of adaptation deeply marked, part of it can be seen on the front part of the pupil as high up as possible, the circle of correlation is clearly seen and larger than that of a pigeon selected for all distances, full lengths and half lengths, the iris is smaller, of a bright rich colour, unspotted, the fifth circle is clearly apparent, or at least can be seen without examining the pigeon in the hand.

N. B. — The eye (figure 29) being totally devoid of the circle of adaptation can be of any of the three classes which I have just enumerated, on condition that you take into account what I am going to say about the size of the pupil, of the circle of correlation, of the iris, and of the fifth circle.

A VERY VALUABLE PIGEON BECOMING A SECOND CLASS PIGEON

A pigeon. that through the result of an accident has someth'ng the matter with the eye, or has any bodily injury must be classed as second rate, bad plumage, resulting from too long a race, or when the pigeon had to remain out in the wet or at night.

The perfect eye must have a very round pupil, the circle of adaptation clearly marked encircling the pupil completely the circle of correlation should be very thin and complete, a large iris of a rich colour, the fifth circle thin and quite visible.

UNSUITABLE PIGEONS.

The pigeon having the eyes badly placed, the pupil deformed, totally or partialy devoid of the circle of adaptation, possessing the other circles according to sizes necessary, but spotted or pigmented or one overlapping the other, the iris being of a pale colour, dead and bloodless. One of these defects is sufficient to depreciate the pigeon.

INTERESTING REMARKS

The eyes play a very important part in the study of pigeons, after having described the points which you must follow when judging the eye for racing and for reproduction I must now draw your attention to the mistakes which are possible to make when judging a pigeon only according to the eyes. Although they show many qualities and defects, it is absolutely necessary not to follow only the indications shown through the eyes. It is necessary to observe the physical moral and constitution, as is described in the first lesson.

It is well know that there is a close relation between the value of the eye and the physical qualities of the pigeon.

Generally the pigeon with perfect eyes has also a perfect body, the muscles of the wings are strong, and the wings well shaped so that it can take flight with case and rapidity.

Figures 26 and 27 have the characteristics of a pigeon of high birth (origin) already described, and the eye is so well set, and defined, that it proves the splendid constitution of the pigeon, the memory, home-coming instinct, will-power and intelligence.

When you want to mate pigeons for reproduction you must give the preference to those birds with the qualities well marked ; pupil of average size, round even, no spots, the circle of adaptation must be complete and thin and the circle of correlation well marked, the iris of a rich colour, free from pigments or any blemish whatever.

The fifth circle must be visible (Never mate with a view to reproduction — types which have not both the circle of adaptation, for if one or the other has not the complete circle you will never have select pigeons as a result of this cross-breeding.

It has often been seen however that many pigeons having this circle incomplete have been honourably and advantageously classed in the racing competitions, but on no account must such pigeons be mated with the view to procreate. You will not have good young ones if it is only from the first two eggs laid or the two last of the season, when the two pigeons possess different characters, forming after some time only one, and even then they must both be descendants of great pigeons (high blooded), that is to say possessiong all the necessary qualities.

If you possess such a pigeon, enter it for the racing competitions as long as it comes out with success, but do not keep it for reproduction. Take note also when mating pigeons with eyes such as figure 29 represents, totaly devold of the circle of adaptation — a union between two such pigeons cannot produce good birds, from the first two eggs laid or the last two perhaps when the two characters have become more similar.

A pigeon with eyes as represented in figure 27 (full eye), although it is totally devoid of the circle of correlation but having the circle of adaptation complete the iris of a bright rich colour, occuping the ouside zone, suits when mated with a pigeon having eyes as represented in figures 26, 28 and 29, on condition that you follow instructions given about the last two figures.

Mate only birds with perfect eyes, if you want good racing pigeons. Don't forget that all colours are good on condition that the eyes have the other qualities described above.

I have spoken about the qualifications of the pigeon and how to find them out because I wish to make known to you something of the greatest importance, and something necesary for the next chapter which treats of mating for reproduction ; and also of mating racers.

A fancier must study the eyes thoroughly if he wishes to have good results.

By observing the rules which I have described, you will obtain first rate pigeons, for racing as well as for reproduction.

Many fanciers who followed my advice with regard to the eyes and mating told me of the wonderful results they obtained. They, did what they saw me do during the public meetings and also at my own residence.

The excellence of this method has been proved many times.

Selecting and Mating the pigeons

In pigeon-breeding selecting in order to carry off prizes means, keeping only first class pigeons for racing and reproduction.

Selection is a sort of magic wand, which brings everything under its controle.

Without selection, you can have no real type no improvement in the breed, nor progress in the evolution, and no successful result. Pigeons left to themselves, that is to say left to mate as they like, will only ruin the fancier. When you want to begin pigeon-breeding, or to better your colony the first thing to do is to choose pigeons of the best strain, you must not attach great importance to whether they were successful in the competitions or not.

This theory may seen very strange to you Indeed but you must become an adherent of it.

For this purpose I think it preferable to let you know about the laws of Mendel, which I have adhered to for several years.

It was principally on plants that this great scholar experimented. Since his discovery the same experiments have been effected with animals and the results obtained in America and in England leave no doubt as to their value.

Nature does not separate the animal kingdom from the plant kingdom, in respect to the creation of forms, growths and evolution.

You need not be afraid to apply the laws of Mendel in trying to improve your stock.

The general selection is divided into two parts i. e. artificial and natural ; the aim of the artificial selection is (taking into account the natural selection and its governing laws) to improve or alter the breed by judicious mating : the second selection is to try to continue the strain which is best adapted to the surroundings and conditions of life to which they are subject.

OUR ANCESTORS' METHOD OF BREEDING

Pigeon breeders of former times mated only pigeons which were good racers and this method naturally resulted only in a terrible loss.

Amongst the large number of young ones the majority were rejected or lost in the training and there were just a few pigeons which resembled the type desired; and that is the reason there were so few good pigeons at that time.

Wendel's system did away with the old method altogether.

Since the principles of consanguinity have been made clear, pigeons-breeders know how to produce good pigeons.

By uniting pigeons according to the qualifications which I pointed out to you about the eyes, and the moral and physical conformation, also the origin of the hen, you will surely and speedily succeed. You can judge from the progress of your operations. You will sacrifice fewer pigeons by selecting them and you will be able to keep up the honour with a very small number of pigeons. Through this method you will also have the advantage of knowing what to do to form a special type possessing all the qualifications for racing and reproduction, while retaining the colour of the plumage you want.

As soon as you have found out all these elements the line of breeding will be definitely traced.

When you have a bird with the right character and another with a thoroughly contrary character you will have no difficulty in deciding what to do.

HOW THE DIFFERENT VARIETIES ARE FORMED

The fundamental colours through which all our present varieties have come were only black, red and white.

The bluish colour comes from black and white ; the yellow, from red and white. This is the reason why the fancier gets different results direct, from two colours alike which nowithstanding does not contradict the law of mendel.

Experiments show :

Blue is composed of white and black, therefore it is a sort of dirty white, if a blue is mated with a dirty white there will be whites and blues. If a blue and a white are mated you can also have blacks and blues, and also when blue is mated with blue ; the production manifests itself in proportion to the number ascertained.

Out of each number of four there are two blues, one black, and one white. In the second generation the whites and blacks remain faithful to their shade; the blues continue giving two blues, a black, and a white. The following table gives the different production :

Black with white gives blue;
Black with black gives black;
White with white gives white;
Blue with blue gives two blues;
Black with blue gives one black and one blue;
White with blue gives one white and one blue :

In the first compleg, the corresponding characters are united, the stability exists as long as the production developes, but if new germinal cells present themselves, the issue of this union will be changed and instead of a bluish plumage, you will have blacks and whites again.

The table of these different colours will therefore constitute the following : if the descendants of blues are united together they will give the type they have inherited themselves. If two germinal cells are in contact with a descendant from white, the issue will be white ; it will be the same case if two blacks be mated but if the union is between a black and a white and one cell remains for each, the result will be two blues ; one black and one white.

The same combinations can be formed with the yellows, a variety direct from red and white.

In this way the law of Mendel, confirm and prove the rules for breeding which all breeders adhere to in uniting different breeds.

In cross-breeding reds dominate over white, and pure white dominates over the other colours.

The young ones from a union of coloured pigeons with whites gives an average of three whites to one coloured one, the whites thus obtained when mated together give coloured descendants.

By scrupulously observing the rules which I have given, you will learn from experience that a certain couple will always give a certain shade, and that a certain plumage always produces in its dominant tone and also a certain colour is generally accompanied by certain qualities.

You will also come to the same conclusions about the eyes, the moral and physical conformation, you must take into account and consider the minutest details I have stated, which will prove to be of the greatest importance.

By studying the rules which apply to the colouring of the eyes and the plumage, you will obtain about the same results every year, and you will obtain results which will keep you on the right road to success.

The hen destined for reproduction must be protected against the influence of all cocks except her own, and both types must be as close as possible; the hen is always more slender than the cock.

In order to train your group in a favourable evolution, it is necessary to keep the sires a part until the second egg has been laid. You can sometimes make a mistake in the progeny, if your pigeons live pell-mell or out of your sight.

MATING BETWEEN BLOOD RELATIONS

The aim of reproduction is to have new types for racing of for further reproduction. It will therefore be in your interest to mate only ideal pigeons through which you can form a select pigeon-loft : they should be health, well formed, free from any hereditary defects, descendants of high breeds, the parents must be strong, and able to combat all difficulties and thus win the admiration of all adepts of your favourite sport. I should like to draw your attention especially to the necessity of effecting unions between perfect birds which resemble each other as much as possible generally the hen is more slender than the cock ; the colouring however can differ, as there are only three different colours which form the following varieties :

1) Dark reds ; tan, grey, lead-coloured ; dark grey-coloured ;

2) Red scales (shell), silvery, bright scaled; mosaic; chubs;

3) Whites, blues ; pale or light colours, the wings of which have white feathers.

Althrough I insist that the two procreators possess the same characteristics, an exception can be made for unions between pigeons with the « full » eye, unless you wish to have those which are classed in the first rank for reproduction. Consanguinity is the best principle by which you arrive at a direct result but it is a « double-edged » weapon, very dangerous to manipulate. In order to obtain a group of splendid pigeons this « weapon » should be employed on good grounds and the instructions which I have given carefully carried out.

Only pigeons of a perfect conformation possessing all the necessary qualities and the eyes (fig. 26, 27, and 28.) can be devoted to reproduction when they are closely connected.

1st EXAMPLE

Extra bird : Perfect eyes with perfect eyes; both having the circle of correlation, and the iris of the same colour, and having the plumage of the same sort or quality.

2nd EXAMPLE

Extra : Perfect eyes with perfect eyes, one having the circle of correlation yellow and the other having it white, both with the iris of the same shade and the plumage the same.

3rd EXAMPLE

Extra : Perfect eye with perfect eye both possessing the circle of correlation the same colour, the iris of a different colour but of a bright rich shade, the same plumage.

4th EXAMPLE

Sire : Full eye with perfect eye, circle of correlation yellow or white quite visible, totally encircling the circle of adaptation, both having the same coloured iris, of a rich colour, and same quality of plumage.

5th EXAMPLE

Sire : Full eye with perfect eye; circle of correlation yellow or white, iris opposite colours, rich and brilliant, same plumage.

6th EXAMPLE

Extra pigeon : Perfect or full eye, with racing eye, possessing a third of half the circle of adaptation, circle of correlation and iris of the same rich brilliant colour plumage same quality.

In this example although one of the pair has not the circle adaptation complete but possesses a clear circle of correlation with the iris and fifth circle of a bright rich colour it can be connected with pigeons of the same family, having eyes as in figures 26 and 27 when you want to reestablish a branch which has become almost extinct.

From the particulars which I have given you, it should be quite easy to choose pigeons for mating on condition that you possess pigeons with those qualities and that you take into account varieties of the same class.

CROSS-BREEDING

The idea of cross-breeding is to try to remedy the defects in one or other of your pigeons or to prevent them from developing in any of the descendants. To arrive at a good solution you must first study respectively each pigeon's pedigree as far back as possible observing closely their physical and moral conformation : it would not do if in trying to remedy or avoid certain defects you introduce others.

If you have pigeons which are too light, too thin, or too heavy, which have great difficulty in coming into form during the racing season, you will never obtain good results by mating two such pigeons.

If you have others which are too low or two high on their feet, or with too long or too short a beak, you will never obtain first class birds from such unions.

When you are coupling for reproduction with the view to re-establish a group of valuable pigeons, you must work carefully and gradually, otherwise you will spoil all the qualities which the birds possess and you will obtain defective young ones.

If you want to correct a defect, choose a pigeon of the same blood, resembling as closely as possible the one that has the blemish which you want to wipe out.

Nature has only one end in view to better the defects ; and in the long run she wipes them out altogether. If you try to remedy several together, there are ten chances to one that you will obtain a lot of good-for-nothings.

In your operations, always try to develope the necessary qualities ; strength, endurance, intelligence, and memory. Very often cross-breeding is effected without the least idea of the subjects, and generally you have only very poor pigeons as a result, or as we say just « eaters of grain ».

To arrive at a successful result, be careful to choose only capable sires, i. e. perfect pigeons extra from every point of view, enjoying perfect health, middle aged, eyes as shown in figures 26 and 27.

Working in such a way, you will have young ones of incomparable beauty, fit to make you a prize-holder in the grand international races.

Do not forget that a perfect bird mated with another possessing the same qualifications and of the same family, will produce a perfect pigeon.

Then, if a perfect bird is mated with a good racer (figure 28) of the same family, the progeny will inherit the defects of the latter which will be four times as great in the first generation and so on resulting in a great many combinations of several different types.

If you mate pigeons too young you will obtain types which can be classed for short tosses. However when you have a cock or a hen, old enough, which gives you young ones with many qualities for racing or for reproduction, it is a good thing to couple it with a younger one.

Concerning young birds resulting from pigeons that are too old you risk obtaining late types, that is to say, pigeons which will deveope too late, which will have difficulty in coming into form, and this, perhaps at the end of the racing season.

The best age for the two pigeons is from two to four years. The conditions are the same for kock and hen.

If your loft contains only perfect pigeons, that is to say, extra from every point of view, mating will only be child's play.

Unfortunately those white « Blackbirds » are in the minority with nearly every fancier.

You must remember that our pigeons descend from three colours, blacks, reds, and whites, and that these colours have given us all the present day varieties and there — fore a thorough knowledge of mating becomes more and more complicated and important.

All these varieties should therefore suit the three distinct colours, and should be of the same class according to the colours cited above.

The way in which you mate your pigeons is very important for it is on that the result depends.

First of all the conditions should be to choose as producers, two perfect pigeons, classed as first rate.

I must also draw your attention again to the fact that with the view to reproduction you should mate pigeons having as close a resemblnce to each other as possible.

The same sort of plumage and the same corpulence not forgetting of ocurse that the hen is always slighter than the cock.

If possible you should find out the physical and moral characteristics of their ancestors so as to make a union between types, the parents of which, gave every proof of their abilities.

Mated according to these conditions, quite a number of good pigeons should be produced which could be classed in the first rank.

Of course it is not possible that the whole family inherits all the qualities of their procreators, there will always exist some pigeons devoid of the brilliant qualities of their parents. Even if you know every little detail in their ancestors, back to a very remote period, the defects also come from some ancient type : the Persian messenger to the carrier pigeon.

Experience has taught me that generally the first couple of eggs gives the maximum number of qualities, or all that one hopes to come from a judicious mating.

When a union has been properly effected ; according to the points which I have described, there will be a fair number of good pigeons amongst the descendants coming from the different pairs of eggs that is to say they may be as good coming from the second, third, or even fourth pair as from the first pair.

Only I must once more point out to you an important fact ; if you have only one good pigeon from this union, it will be from one of the first pair of eggs laid.

It will be quite easy for you to see, through this, if the union has been effected under good conditions or not.

If the first couple of eggs give you weak pigeons, put an end to this union; you only lose time and money.

If at first you get a suitable young one and the second eggs give unsuitable birds, you should stop the production; on the contrary, if from the second or the third pair of eggs you obtain valuable pigeons you are on the right road and you have only to continue.

By the first pair of eggs I mean the first two laid 9 days after the union.

The fancier who practises « great » widowhood should separate the cocks and hens after they have been hatching together for ten or twelve days until the competitions are over.

Through this union he gets the first pair of eggs which the pigeons in the reproducers' loft, can rear.

If, from the same hen he wants to have another first pair of eggs, he has to make a new union with a cock from another loft ; then he can continue in this way if he wants several first pairs of eggs, on condition, naturally to choose a valuable cock each time.

At all times, cross-breeding has been necessary to improve the breed ; it has been the basis of evolution from the first stocks to the well-known types of our days.

I again insist on saying that you must not try cross-breeding with pigeons of a stock which, according to certain breeders, was of great renown and are at the moment effected in any way.

In unions made with a view to procreation you should remember that the proper selection of the producers forms the soul of the group, and that is therefore the ground work of your future in pigeon breeding. The best system is to know and select the pigeons which you consider most suitable for yourself after having studied the instructions given in the lesson.

All perfect birds, i. e. extra from every point of view can be mated for procreation.

This class of pigeon — extra from every point of view includes several distinct categories which are :

1) A pigeon possessing qualities to excel in a race of 900 and 1.000 kilometres and also inheriting the qualifications for a reproducer will be classed as perfect ;

2) A pigeon which can be classed with honour in distances from 300 kilometres on, and which can go as far as Spain, and possessing the necessary qualities for a reproducor will also be classed as perfect ;

3) A type possessing the desired qualifications for fast races, from the first race to 300 kilometres and having the qualities for reproduction will also come under the list « classed in every respect ».

Another having the qualities of a good sire or dam, but not of a racer the reason being either the result of an accident or because it has the « full eye », will also be classed as perfect, for reproduction only, exception can be made for the latter which will be judged according to its physical conformation.

The way by which you can make sure of these qualities, will be stated in the next course treating on classification according to my own method :

EXERCISES

1st EXAMPLE

1) Pigeon for reproduction apt in all distances, eye figure 26 :
plumage same colour,
pupil average size,
cercle of correlation yellow,
iris the same colcur, zone average size
fifth cercle visible.

WITH

Pigeon for reproduction, apt in all distances, eye figure 26 :
plumage same colour,
pupil average size,
cercle of correlation yellow,
iris the same colour, zone average size,
fifth circle visible.

2) Pigeon for reproduction, half length races, figure 26 :
plumage same colour,
pupil a little smaller,
circle of correlation yellow,
iris of the same colour, zone more developed,
fifth circle slightly visible.

WITH

Pigeon for reproduction, half length races, eye figure 26 :
plumage same colour,
pupil a little smaller,
circle of correlation yellow,
iris of the same colour, zone more developed,
fifth circle slightly visible.

3) Pigeon for reproduction, short distance races, eye figure 26 :
plumage same colour,
pupil seeming more developed,
circle of correlation yellow,
iris of same colour, zone narrower,
fifth circle quite visible.

WITH

Pigeon for reproduction, short distance races, eye figure 26 :
plumage the same co-pupil seeming more developed,
circle of correlation yellow,
iris of the same colour, zone narrower,
fifth circle quite visible.

4) Pigeon for reproduction and racing, eye fig. 26 :
plumage the same colour,
average pupil, seeming a little smaller or a little bigger,
circle of correlation yellow or white,
Iris of the same colour,
fifth circle visible.

WITH

Pigeon for reproduction, eye fig. 27 :
plumage the same colour,
average pupil, seeming a little smaller or a little bigger,
circle of correlation yellow or white,
iris the same colour,
fifth circle slightly visible.

5) A Pigeon for reproduction and for racing with a type having the qualities in the three first categories just described, taking into account the physical conformation, and that there should be as close a resemblence as possible.

2ⁿᵈ EXAMPLE :

1) Pigeon for reproduction, apt in all distances, eye figure 26 :
plumage same colour,
pupil average,
circle of correlation yellow,
iris same colour, zone average,
fifth circle visible.

WITH

Pigeon for reproduction, apt in all distances, eye figure 26 :
plumage the same colour,
average pupil,
circle of correlation white,
iris the same colour, average zone,
fifth circle visible.

2) Pigeon for reproduction, full length and half length races, eye figure 26 :
plumage the same colour,
pupil seems a little smaller,
circle of correlation yellow,
iris of the same colour, zone developed,
fifth circle slightly visible.

WITH

Pigeon for reproduction, full length and half length races, eye figure 26 :
plumage of the same colour,
pupil a little smaller,
circle of correlation white,
iris of the same colour, developed zone,
fifth circle slightly visible.

3) Pigeon for reproductoin, short distance races, eye fig. 26 :
plumage of same colour,
pupil more developed,
circle of correlation yellow,
iris of the same colour, zone narrower,
fifth circle quite visible.

WITH

Pigeon for reproduction, short distance races, eye figure 26 :
plumage of the same colour,
pupil more developed,
circle of correlation white,
iris of the same colour, zone narrower,
fifth circle quite visible.

3rd EXAMPLE :

1) Pigeon for reproduction, apt in all distances, eye figure 26 :
plumage of same colour, average pupil,
circle of correlation same colour,
iris rich colour, average zone,
fifth circle visible.

WITH

Pigeon for reproduction, apt in all distances, eye figure 26 :
plumage the same colour, average pupil,
circle of correlation same colour,
average iris, different colour, rich,
fifth circle visible.

2) Pigeon for reproduction, full length and half-length races, eye fig. 26 :
plumage of the same colour,
pupil a little smaller,
circle of correlation of the same colour,
iris a rich colour, developed zone,
fifth circle only slightly visible.

WITH

Pigeon for reproduction, full length and half-length races, eye fig. 26 :
plumage the same colour,
pupil a little smaller,
circle of correlation the same colour,
iris different colour, rich, developed zone,
fifth circle only slightly visible.

3) Pigeon for reproduction, short distance races, eye figure 26 :
plumage af the same colour,
pupil a little bigger,
circle of correlation same colour,
iris of a rich colour, zone narrower ;
fifth circle quite visible.

WITH

Pigeon for reproduction, short distance races, eye figure 26 :
plumage same colour,
pupil a little bigger,
circle of correlation of the same colour,
iris different colour, rich, zone narrower,
fifth circle quite visible.

I consider it useful after giving the above examples to point out to you that if you wish to have pigeons with

« full eye » you should mate two pigeons having the « full eye » and in other respects resembling each other as closely as possible.

You will also notice that when mating with the view to a reproduction of this type you can only remedy one defect at a time and the unions must be made only with pigeons classed in every respect eyes figure 26, and 27 and that all other unions will give you only very poor pigeons.

Above all avoid « degeneration ». ! a thing which frequently happens when an inexperienced breeder does not take into account the plumage, etc., of the pigeons he is breeding.

CREATING A NEW STOCK

The creation of a new stock should be effected through the union of two different families of which you know every detail, concerning their ancestors.

Supposing you desire to form a new group of pigeons having the following characteristics, blue feathers showing three black transverse lines, one white feather on the eye in front or at the back; head slightly flattened, forehead wide and slightly arched; on the upper part; average beak, caruncle and eyelids thin not overladen; brown eye or eye with correlation and red iris, physical and moral conformation, etc.

You must choose from two different families two pigeons of which you know every little thing concerning their ancestors without troubling whether they themzelves have raced or not.

They will represent the respective characters of their ancestors, they will be both classed as perfect, having the qualities of good racers and good producers.

. They should resemble each other as closely as possible (except the eyes), one must have the eyes represented in figure 26, and the other those in figure 27; or both can have as in figure 26, the circle of correlation and iris different, remember that all the colours are good.

Those two types being well-known for their qualities, the greater number of the young ones of the first year will be perfect pigeons, possessing all the necessary qualities for good racers and good producers.

In this first generation, it is more than probable you will have some birds with different coloured plumage, eyes, and other characteristics different to those of the father and mother.

So as to mantain this stock and to have success with it you must only keep the pigeons which come nearest to the type which you tried to have.

They must undergo a severe inspection each little point must be studied and any pigeon not possessing the necessary qualities should be discarded.

Then, to obtain good results and to be sure of arriving at success, mating should be continued in consanguinity that is to say between descendants of the same family.

Method. — First of all you must mate two young ones from the first generation, therefore, brother and sister, after having chosen the two best, and possessing most of the qualities of the parents. After having taken away all the pigeons which had not the qualities for reproduction there now remains only the types from two successive generations which resemble each other closely in character and qualifications.

To obtain the third generation you must make your unions in the following way; father with daughter, or mother with son and from this issue you must keep only the young ones which resemble most the father or mother.

You will therefore have in your group three successive generations descending from two different families little by little each birds qualities will blend together and after a short time will leave only the characteristics which you want.

The fourth generation, as well as all the others are obtained by uniting two pigeons which are related to each other outside the second degree, for instance : grandmother with grandson, or grandfather with granddaughter.

Many years experience of this kind have shown me that the family characteristics are easily trnnsmitted by mating closely connected members of a family :

1) Brother and sister, both off-springs of subjects of high origin;

2) Father and daughter, the latter being an off-spring of brother and sister;

3) Grandfather with grandaughter an off-spring of brother and sister or father and daughter.

In these unions the operations are the same for the hen as for the cock.

A very important point is to take note every year of the number of young ones born. their different characterictics, from which parents they are off-springs, for what particuiar reason they have been reformed for reproduction, the number of valuable pigeons, as well as the middling ones, in fact take note of all particulars which will enable you in the future to judge the improvements which take place, or if degenerency results.

The description of each off-spring from these unions should be carefully written in a book which you should always keep near the loft.

All the particulars noted will enable you to see which characteristics dominate and which are weakest in your new stock. By the preceding examples, and with a knowledge of the laws of Mendel your researches will be simplified and you will know exactly what to do in subsequent operations.

After serious study and a severe selection you will soon discover if your unions have been operated under good conditions, according to the number of pigeons you will be able to keep as reproducers from generation to generation.

It your choice in procreating pigeons has been made on good grounds and that you have scrupulously followed the method which I have taught you, at least 75 % of the descendants from the fifth generation should be of the type you want.

On the contrary, if a mistake has been made, the defects will increase more and more and will result in complete degeneracy.

The first indications of degeneracy is generally noticed in the feathers and eyes, the latter will become deformed, the circle of adaptation spread over the pupil, the iris spotted, etc., etc.

RACING COUPLES

This is a very important point, and I find it my duty to tell you how you should mate your pigeons so as to win prizes each year and to remain on the right road.

When the « marriage » has been arranged between the right pigeons, the cock will not accept advances from any other hen but his own; and the hen will reject any attentions paid to her by another admirer.

To mate or marry pigeons is not difficult, but to find for the cock a hen which he likes is a very difficult problem and one which few fanciers can solve, although it is of such importance.

This is the only course on mating, which will give you such information, I am sure.

It is necessary, so as to arrive at this solution to place all the different elements at your disposal in order to leave as little as possible unforeseen.

A marriage made with the view to procreate does not always suit for racing pigeons, it can be arranged between pigeons, one of which, or perhaps the two are destined to take part in the competitions.

The chief thing in such a case is to give the cock a hen which will make hen happy; and for which he will put forth all his efforts when competing, the same is to be said of the hen when it is she that competes in the race.

The plumage of the hen must be an opposing colour to that of the cock, if the cock is an off-spring of a judicious cross-breeding with plumage all the same colour and the eyes of a different colour, if it is a descendant of a pair which were near blood relations.

My long experience has proved that an off-spring from birds of a « judicious » mating, is preferable to that of a pair, one of which, has different coloured plumage and that a young one whose parents are related in blood always prefers a member of this family with the same coloured feathers, the eye of a different colour.

Therefore when you make unions between members of the same family with a view to racing, be sure that one of the pair has a different coloured circle of correlation or iris, taking into account the different varieties which form part of the same rank or class, and which as I have already said are produced through three distinct colours red, black and white.

When a pigeon is suitable for short distances, if it is an off-spring by sanguinity you should mate it with a type possessing qualities for long distance races, the colour of the plumage should be the same on condition that the iris is larger and of a different colour.

You should do just the contrary with a pigeon racing in long distances and an off-spring of parents closely connected, you must give it a mate destined for fast races the colour of the plumage should be the same, the circle of correlation larger and the iris narrower and of a different colour.

To a type of pigeon suitable for fast races and a descendant of a « judicious » mating you must give a pigeon with different coloured feathers trained to compete in long distance races.

To another type having qualities for long distance races and a descendant from a judicious mating, you must give a partner with feathers of a different colour and trained to take part in fast races.

The conditions which I have just stated to you are very rarely practised; most of the time they are practised unawares by a few fanciers who discover it after having learned the method which I have just described. Judge for yourself and study your mated couples which distinguish themselves in the competitions and you will agree with me.

You can prove this when you want to mate your pigeons before starting for the races, and when the hen is in the nest by giving it a first rate cock; if is an off-spring of blood-relations it will choose a hen belonging to its own family with the same coloured feathers (taking into account certain changes) the plumage has undergone and if it is a descendant of a « judicious » union it will choose a hen of a different colour.

An earnest pigeon-breeder will therefore be just as careful about the unions for the races, as for reproduction ; by following these instructions he is sure to prosper in pigeon-breeding.

In my indications concerning the colour of the feathers, the eyes and the form of the body I have pointed out that there always exists a dominant characteristic in one or other of the two pigeons.

It is these opposing characteristics and qualities energy and will-power, love, endurance, create speed, which is the most essential thing when returning from the races.

If you make a union between two pigeons having a tenacious character, or strong will-power (with the view to racing) you will never have harmony between them, one of the pair must always be a little smaller, or lighter so that it can fly about quickly from corner to corner in the loft in order to excite its mate to jealousy.

These are most important factors.

To let you, know how to operate unions for racing. I have made out the following table showing you how to act under good conditions, and to have birds that will enter the loft quickly and fly fast during the races.

DESCENDANTS FROM CROSS-BREEDING IN THE SAME FAMILY

Same plumage with same plumage.

Varieties of same rank, with varieties of same rank.

Short distances with long distances.

Yellow circle of correlation with white circle of correlation.

Narrower iris with larger one.

Long distance with short distance.

White circle of correlation with yellow circle of correlation.

Developed iris with narrower one.

Full eye with developed pupil.

DESCENDANTS FROM JUDICIOUS CROSS-BREEDING

Long distance with short distance.
Light plumage with dark plumage.

Yellow circle of correlation with white circle of correlation.

Different iris with different iris.

Full eye with developed pupil.

Short distance with long distance.

Dark plumage with light plumage.

White circle of correlation with yellow circle of correlation.

Different iris with different iris.

Developed pupil with full eye.

I specially insist on the importance of this process of mating for the racers, many times I have seen, and remarked to fanciers that a certain pigeon had the qualifications to excel in all weathers; but on account of being unsuitably mated it could not be classed even once all the season. Afterwards those breeders told me I was quite right as the following years when the gave that pigeon another hen, it distinguished itself during the whole racing season.

After that they always gave the cock the hen it liked, and from the beginning of the competitions until the end, during all kinds of weather, the cock was triumphant.

At the commencement of the season when you begin mating for racers if you breed young ones from these racers let them hatch the first two eggs laid, generally the first eggs are laid nine days after the pigeons have been mated.

HOW TO CONSTITUTE THE COUPLES (PAIRS)

The pigeons which are united must be carefully chosen according to their aptness and qualifications vis a vis with each other, a pigeon full of vigour will not tolerate another one, especially if it is closed up in its pigeon hole.

The pigeon-holes such as I have described in the chapter about lofts, must have a partition, so that it can be divided into two parts, thus preventing the pigeons from fighting or pecking each other. The hen is to be in one compartment and the cock in the other, the latter should be put, the side where the nest is.

The partition which separates them being in netting or lattice will enable them to see each other, in the beginning probably one of the pigeons will not show any affection for the other but will try to get away, it is only little by little that they will become attached to each other.

The cock occupying the section where the nest is placed will go and sit in it calling the hen from time to time by raising its head; sometimes it will go to the partition cooing, and dragging its tail wide open along the ground.

The hen will approach, and also go about dragging its tail on the ground.

When this has continued for some time the two pigeons will begin to caress each other with their beaks, and come closer to the partition at this moment you must raise the partition, and your pigeons are mated and will both occupy the nest without further trouble.

As a precaution and to be sure of your breed, you must keep the hen in its own pigeon-hole until she has laid the second egg; let them make their flights separately, first the cocks and then the hens.

This is of the greatest importance when you are mating for reproduction, the result of a hen being courted by another cock can spoil everything.

LATE SEASON MATING

Many fanciers, practising great widowhood have the habit of mating their pigeon's again as soon as the competitions are over, so as to have a few more young ones, thinking that the late birds have better qualities and are more apt to become famous than those born at the beginning of the season or at any other time of the year.

During my many years experience I have always noticed that when the parents had to endure a whole season of difficulties, and had been overworked during bad races, they were from being able to produce capable birds, neither through sanguinity nor judicious mating.

Only, if these racers had been kept and well managed, they could have produced very good youngsters, for the reason that having been subjected to « great widowhood » they would have been separated all the racing season and the pigeons born in the late season would be considered as having come out of the first couple of eggs laid, especially if the union had been made between members of the same family and that the producers had the same sort of plumage.

On the contrary if these young ones were from a judicious mating, and if the parents had been kept back or not, you can obtain only one valuable pigeon, both pigeons not having the same characteristics can only give one type resembling only one of the parents — the dominant one.

What made some fanciers suppose that there is a much better chance of obtaining good pigeons in the late season is because many pigeon-breeders mate their pigeon much too early; the young ones suffer in the cold weather, some of them even perish from the severe temperature and many pigeons from the first couple of eggs die.

When it is a judicious mating, bear in mind that it is always from the first couple of eggs that a good pigeon comes, only it often happens that when the birds are mated very early a number of pigeons which could be classed in the first rank, cannot give proof of their qualities or aptness.

This cannot be said of late breds, they have fine weather, the young ones benefit very much as they are generally kept from competing the first year, so they have time to develope properly and be in good form for the next year, especially if they are reserved until they are two years old, a thing which most fanciers do.

According to the methods which I have taught you in each chapter about the management of your pigeons for racing, if you mate your birds only from the mouth of march on, you will have much more chance of obtaining a large number of superior pigeons, than if you wait until the bate season, because the parents is the first place will be healthier, and well rested; during the winter season the blood has been purified.

When in the different competitions of the season there has been one or two very fatiguing races, and some of the pigeons had to remain out at night, or that they had to be kept several days in the baskets, which weakens the pigeons dreadfully, it sometimes happens that the fancier is obliged to make late season unions to replace the missing pigeons. This prevents him from introducing a new breed into his loft.

To remedy this you must be careful to mate only pigeons which are well disposed, that is to say ones that have not raced too much or have not reared too many young ones.

Bearing in mind that heredity transmits all the defects of the parents to the off-springs, therefore you must mate first class pigeons only.

Before finishing this question on mating I consider it necessary to point out to you that you must proceed slowly and gradually, if you have fast racing pigeons, retain this precious quality, by cultivating in this line, it is the progressive work of each union which gives us pigeons fit for long and difficult races.

Management
of the loft for great Widowhood

The loft destined for total widowhood can only be occupied by pigeons of the same sex, generally it is the cocks which are reserved, for the reason that they can bear separation from the hens more easily.

This loft must not have more free holes than there are pigeons, they must be as far away as possible from the other lofts.

It must be constructed according to indications in the lesson about lofts, it must be facing south or south-east, with windows, skylights, or glass tiles, so as to get as much light as possible this being as necessary for pigeons as for man.

There must be plenty of sunlight especially.

Through the rays of light and heat other rays are emitted which bear a great influence on all living things and beings.

What do we see in human life ?

Children suffering from rickets or any other disease being cured through the ultra-violet rays. Those wonderful rays are excellent for the health of our pigeons, and at the same time they kill all microbes and destroy the eggs of other insects, in fact they give new vitality to the pigeons.

For this reason your loft must be well situated and constructed so as to get as much light as possible.

Pigeons playing the role of « great » widowhood should be alone during all the racing season, they must be mated at the beginning of the season just the same as the pigeons in the other lofts. The mating for all pigeons is generally at the same time of the year, the cocks must get the hens they prefer, but with just the opposite charasteristics, for the reason that the racing pigeon has to be made jealous of its mate.

In the lesson on mating I shall give you other details, you have already learned that a pigeon mated with a certain hen distinguishes itself in the races, whilst when mated with another, cannot succeed in being classed.

The system of « great » widowhood « does not allow a pigeon to rear a young one either at the beginning or at the end of the races.

Some fanciers say that in order to be in good health the pigeon must rear young ones, others say that it must have lost its first flight feather before being put into the basket.

However the results which I have obtained through my several years experience have proved the contrary, the best system for any fancier is to remain near the loft and observe the pigeons, while performing their daily exercises.

A pigeon which has been breeding has weaker muscles, spoilt caroncules, it moults much earlier and therefore cannot take part in the competitions at the beginning of the season.

If the weather is fine, pigeons playing « great » widowhood should be mated about 4 weeks before the first training exercises begin, i. e. the first of March if your pigeons are healthy, the first egg will be laid the 9th day.

Let the parents sit on the eggs for ten days, then take away the hen, and leave the cock for a day or two longer, he will come away of his own free will. When the hen and the eggs have been taken away, the cock will remain alone in its hole; this proceeding will not raise their spirits, but you will notice that seperating the sexes will cause one or other of the pigeons an amount of suffering.

You will even see that at such a time athough the pigeon receives just the quantity of food which is strictly necessary, it will still leave some grains in the trough.

It is therefore quite natural to give the pigeon some time to get back its moral and physical forces again, this is done either by stopping the daily exercise, or making it take a rest after the evening meal when you must make it dark inside the loft. I have already said that pigeons playing in « great » widowhood must never rear a young one. It is a mistake to think that a pigeon which has reared a young one is more attached to the loft than another there is plenty of proof that cocks which never had a hen were victorious in the competitions.

A pigeon which has not hatched a young one retains all its strength, does not spoil its carancules, has hard muscles and does not moult just in the middle of the races, stays a long time in form and distinguishes itself in the races during many years.

In order to produce these wonderful results « great » widowhood should be practised for a long time, and it is for this reason that, this particular system of the hobby is often abandoned by fanciers just before it can bear fruit.

The hens belonging to these racing cocks after they have been hatching for about ten days, must be taken away during all the racing season to a spot where the cocks can neither see nor hear them.

This is extremely important for if the cocks discover the place where the hens are, you will get no good out of them for the rest of the racing season, and you will not even be able to let them take their morning and evening exercises.

Each time they go out they will fly over the roof of the loft where the hen is, on returning to their pigeon house they will fly about from one corner to another, and after a time will become thin and weak.

The hens belonging to these cock which had taken part in the training exercises, either the year before or at the beginning of the season although kept in seclusion during the tosses, through this method can take part in the races.

If for some reason or another the cock cannot participate the hen can replace it in a short distance races, on condition that she made some preparatory flights the week before she was put into the basket.

When she returns she may not stay in this pigeon-house, she must be put away or into a basket during a certain time with the cock.

Te fancier who practises the method of « great » widow-hood watches his pigeons the whole year round ; he never leaves them to themselves.

He gives them a proper diet according to the season of the year, he takes them from the fields; gives only certain grain, drink, etc., according to the instructions given· for the racers in this loft.

The pigeon breeder must have several lofts for breeding; and for weaning the young ones; for the lone hens, and for the cocks destined for the races. « Great » widowhood consists in leaving a pigeon-hole for each pigeon, there must be no perches. If there are too many pigeon-holes they must be closed up; there must be a nest in each hole that is occupied. The pigeons in this loft must have each, its own pigeon-hole.

Contrary to the pigeons in demi-widowhood it does not matter about the distance of the races, or if they be fast races, full lengths or half- lengths, they must race the whole season without seeing their hens, neither before nor after the tosses.

When there is an important competition at the end of the season, in which you desire to take part, if you have a pigeon which treats a young one favourably you must do identically the same as with pigeons playing in demi-widowhood.

This loft only contains cocks, but if you notice that two of them go together, take one of them away for a few days, put it into a large place where it can fly about and keep its from. When you put it back in its loft take the other cock out for a few days, and continue this process until they do not go together any more.

Real widowhood when properly managed is the system of the sport which brings the highest glory to all our great prize winners.

A fancier who practises this system with perseverance and determination will have no difficulty whatsoever.

The loft containing cocks only more easily kept clean than the loft in which the Natural is practised.

When you have pigeons hatching in the loft for the Natural, very often the hens will just choose the moment when the loft has been cleaned, to come out and dirty it again.

Then during your absence these droppings are spread about the place, and when mixed with the grain can bring on sickness and disease.

A careful breeder will soon find a remedy; during the training exercises morning and evening he will sweep up all the droppings on a little shovel, and in two minutes the place is clean again.

These training exercises take place twice a day, in the morning between 8 and 9 o'clock and between 4 o'clock and 5 o'clock in the afternoon, such hours are only for when the days are long and bright.

The first day the exercises must last only about half-an-hour in the morning and half-an-hour in the evening; the time can be increased every day until you arrive at one hour's exercise morning and evening. These two hour's exercise must be continued during all the racing season.

After having let your pigeons out you must pull down the blind then you must hang a flag outside so that the pigeons will not remain on the roof. For further particulars look up instructions given in the second lesson on training.

While the pigeons are doing their little bit of exercise you should clean out the loft, all the material inside, open the windows, doors, skylights, etc., so as to give it a thorough airing you need not fear draughts as all the pigeons are out.

The moment the loft is cleaned, you must clean the drinking vessel and put fresh water into it.

Then into each little trough; in every pigeon hole — put about 10 or 20 grammes of grain as is indicated in the second lesson on « The proper feeding for this system of the Sport ».

As soon as the time of the flight has expired, take down the flag and open up the entrance again.

Then go near the pigeon holes and as each pigeon goes in, talk to it aud rub it down : this will make the birds familiar, they will soon come up to you without being afraid and without the least difficulty. You should repeat this after each exercise for two or three days. Then put your finger on each pigeon's leg rub it up and down as for as tthe toes. The next day do the same thing, but this time place your hand on the ring ; accustomed to such little kindnesses on returning from the races, they will not be afraid when they see you in the loft, and will expect you to talk to them and rub them down, so you can take off the ring without handling them, and ascertain many little détails without difficulty.

The widowers should be at liberty in the loft from their morning's exercise until that of the evening.

When they have finished their feed after the evening's exercise, and have had sufficient time to drink, the pigeon-house should be made dark inside until the training exercises begin next day.

If one of your pigeons remains thin, and does not keep up with its companions during the flight you should go to the loft early in the morning and examine the droppings, you can see if there is something the matter with the pigeon.

The linen blinds. or shutters with which the pigeon house is supplied to keep out the light will also keep out the strong heat in summer.

The entrance hole and the ventilation hole must always remain open.

In the second lesson, on « lofts for widowhood », I have given the dimensions of the pigeon-holes, and I have also said that they should be divided into two parts by a partition half lattice or netting and half wood. During all the racing season those holes must not be seperated that is to say the occupant must be able to go from one side to the other of this partition.

The feeding trough should be placed in the front. Behind the wooden part of the partition there must be a nest for when the pigeons mate at the beginning of the season, it may be left there so that the widower can rest when it likes.

It is in this nest that certain little operations will be carried on if you have a pigeon which shows good dispositions with a young one or with the eggs, when there is an important competition coming off, you can also practise the same system with semi-widowers (see instructions on widowers) :

If the widowed pigeons do not remain in the nest, (contrary to what they should do in Natural) they will never have good dispositions. This is the point which will prove if the pigeon is in good form or not.

If you have time during the day you should go quietly to the loft now and again to see what the pigeon is doing.

Watch the pigeons through the pane of glass in the entrance door, you will be able to judge exactly if the pigeon is coming well into form or not.

A widowed-pigeon possessing good dispositions, when in its pigeon-hole will be gay, and lively, will look about all the pigeon-house cooing very low from time to time, whilst standing on one foot; it will lie on its nest with its wings a little spread out, and the bill must be closed, and no feathers standing out in front.

I have said that these pigeons must not rear young ones, they must be mated about the 1st of March, they have to be separated from the hen during all the racing season; and they must on no account see the latter either before or after the race.

Pigeons subjected to this system, can be mated again when the competitions are over, but only with the hen for which they have gained all their success during good weather and bad.

They will hatch again for about ten days after which the eggs must be taken away, as well as the nest, holes, etc. The material must be put away in a very cold place but sheltered from the rain, the pigeon-holes should be replaced by little cubes as already cited, about 40 centimetres from each other.

If the hens are in good health they may remain in the pigeon-house until the end of December, but nothing must be allowed to remain in the loft with which a nest can be made.

During this period, which is considered as vacation, on fine days, and between the feeds, the cocks and hens should be out in the open-air.

So as to bring your pigeons into good condition for widowhood, it is necessary to follow carefully the instructions given on training, feeding, and lofts.

LOFT FOR SEQUESTERED HENS

The hen of the « great » widowers must be in this loft, it should be as far away as possible from the other loft ; and built according to the conditions, hygienic and otherwise already stated.

This loft although it does not contain very valuable pigeons should be situated in a light, airy spot. There must be no outlet, but the front must be made of wire-netting, and have a blind or mechanical shutter so that it can be pulled down to prevent the males from seeing the hens when out for their daily exercise. Although those pigeons stay but a short time in the loft for « great » widowhood, they must be kept there under healthy and proper conditions as long as the cocks give you satisfaction.

In the chapter on mating I have given the reasons why the partner of a pigeon which comes out with distinction in the races, should he isolated. If the hens were in another aviary, or in a large loft where there is plenty of air and light, they could mate with other cocks or rear the young ones belonging to the producers, during the races.

The hens in semi-widowhood should take a flight at least once a day.

They can be engaged for the competitions just as well as the cocks, as long as they give good results. A hen in semi-widowhood when isolated will show her « coming into form » as well as a cock, look up particulars already cited on lofts for widowhood.

A hen in good condition cooes like a cock when she sees you and will stoop down when you rub your hand across her back. During the training exercises she will fly round and round in the air clapping her wings.

CONTENTS

THE
FOUR SEASONS

REAL COURSE ABOUT PIGEONS

Published by

M. Joseph HEUSKIN

EXPERT IN PIGEON-BREEDING

Léopold Street, 88, FLÉMALLE-GRANDE - Belgium

Translated from French by Aug. LEMMENS

Course in four lessons, treating of the instructions about pigeon-house managing.

FOURTH LESSON

Training and getting in form — Classement by my method — Diseases and Accidents — Secrets and full inquirements — Method of pigeon-house for the speed concours — Exercises.

Course, Number 4522

Breeding and Training

The purpose of breeding and training is to compel the pigeons to use their sense of direction, and to travel long distances without getting tired.

This training differs according to the age of the pigeon, and the distance to be covered. The same is required from any being, having to accomplish a certain effort. Nobody could suddenly give a maximum of result without previous training.

What happens when a workman stops work for a time ?

When he starts working again, he is tired and even exhausted.

Are race-horses run without training ? In order to win, they must be prepared and trained daily.

All pigeons which are intended for racing, the yearlings included, must be exercised twice a day : in the morning and in the evening, and if possible, at regular intervals before the meals.

These flights in the beginning will be effected with a great care, and for the yearling must coincide with the growth of the fifth feather. It is to begin with a flight of a few minutes, increasing gradually when the pigeons show that they realize what is expected from them, and that their organism is fitted for these efforts. They must not be allowed to overtax their strength. The trainer must be able to judge what he can do without overtaxing his pigeons.

As a rule, I think that the best progress can be made through the following method.

This method must be practised with great care ; mainly for the duration of the flight.

The first morning, the pigeons are feed, released and compelled to remain in the air for about half an hour : Should they try to come down on a roof, they must be chased away, for they must have thirty minutes uninterrupted flight.

An easy way of obtaining good results, and at the same time, to train them to go quickly to the pigeon-house, is to practise what is called « flag flying ». These flags should be of different colours : red, white and blue or any other colours. Any rags can be used, but never use the same one more than once.

Should the same flag be used, the pigeons would soon land even on the flag.

Of course, access to the pigeon-house must be prohibited during all the flight, and from the first day on, when the pigeons have flown for half an hour, you must withdraw the flag, while the usual whistle or call before the meal is given. This call can be given by ringing a bell, or by making any other noise. The usual one is the whistle of the breeder.)

As soon as the pigeons have got into the pigeon-house and can't come out again, let them feed properly without taking any notice of the bird which have not come back and which will be allowed to come in, (taking care not to let the others out), only, when there is no food left.

If you do this, your pigeons will always come home immediately.

This first flight must take place at about half past eight in the morning, so as to have the birds back at about nine ; this being the proper time for the daily feed.

Accordingly to my experience, it is the usual time the pigeons get back from their race. They will acquire a habit of this, and your birds will never be hungry before nine.

In the afternoon, at 4 1/2 PM. in order to have the pigeons back at about 5 o'clock, the same process will be repeated, this being the program for the first day. The day after, or second day of training, proceed in the same way in the morning and at night ; but lengthen the duration of the flight by five to ten minutes, according to the strength of your birds, for as I said, you must never exagerate.

On the third day, proceed in the same manner, but increasing the time of the flight by five to ten minutes.

A flight of about one hour will be obtained and this duration will be maintained but not exceeded, during all the racing season, in any weathers.-

Let it be said, that apart from the flighing hours stated, the birds should not be free.

However when it is snowing heavily or on foggy days ans also during a storm, no pigeons should be set free.

Sometimes, in the beginning, trouble is met with, when having the birds on the wing for one hour at a time.

There are birds which do not fly in a group, but leave the bunch, to go and rest on a roof near by, and the others might follow their example.

As soon as you notice a bird landing, shake your flag. If this has no effect, throw little pieces of clay at it.

Carefully observe this bird to avoid any fresh outbreak of laziness.

After a few days, all difficulties, offered by the position of the pigeon-house, will be overcome and the birds, treated according to my method will fly in a group, and will remain so until the time has come to feed, when the flag is lowered, and the access to the pigeon-house is made free.

Pigeons during their training will be well fed and kept fit, so as to be able to support the training which will do much good to the developement of their muscles, will eliminate all superfluous fat and vivily the tissues by dilating the air pouches (sacks).

If your birds are not fit, you will have much trouble ; but if they are in the pink of condition, you will experience only joy.

Every man practising a certain sport and wishing to beat a record must train methodically and regularly.

If this is true for men, it is also true for animals and specially for the pigeon, the capabilities of wich are increased by severe training.

To be carried out conveniently, the training must follow the four hereafter given rules.

1) The yearling pigeons.

2) The one year old pigeons.

3) Te 2 years old pigeons.

4) Older pigeons.

YEARLING PIGEONS

For these birds, the final goal is not not the same as for the more aged pigeons.

As soon as they are flegdged, they begin their training of their own.

Every day they fly out, increasing the range of their flight daily, and getting well used to the neighbourhood.

The young pigeon must be trained since the year of its birth, in order to develope capacities of finding the cardinal points, varying with each case, certain birds being more precocious than others.

Certain birds are met with, which after having obtained poor results as yearlings, become, when 2 or 3 years old, regular racers, getting prices in every competition and in any weather.

As a rule, a young bird must have grown five primary feathers on each side, before being trained.

It must, before being entered into a competition, be acquainted with the uncertainty of our climate, for while travelling it will have to fight against rain, wind, great heat, etc.

When training, the sentiment of regularity must be awakened in young birds, and one will act wisely in releasing them the first time in company with a few older birds, and only from moderate distances.

These little distances will be covered several times, in suitable weather, taking various starting points, in order to make clear to the young birds, that coming back to the pigeon-house is not always an easy task, and that they must not always let themselves be brought several miles out of their track, while getting used to bad weather.

After having been released several times in group from various distances, they will be released **separately** this time, from the point where the first travel started.

By *separately*, we mean at intervals such, that they do not rejoin in the air, and thus are obliged to come back by their own means.

All the starting points used in group flights will be used once more for individual flights.

Left to their own device, the pigeons will encrease their accuracy in finding the cardinal points.

These training stages for yearlings will be fixed as follows :

The first one about 1 Km from the pigeon-house. (1)

The second one about 3 Km from the pigeon-house.

The third one about 5 Km from the pigeon-house.

The fourth one about 10 Km from the pigeon-house.

The fifth one about 15 Km from the pigeon-house.

The sixth one about 20 Km from the pigeon-house.

The 20 Km, stages = 21860 yards will be adopted each time before a race, as individual training for each bird.

It will be repeated on the day or the day before putting into the baskets, and this will greatly help a quick entering in the pigeon-house.

You will impress in the memory of your pigeons the knowlegde of a flight line only in the neighbourhood of the pigeon-house , it is true ; but this line will be known in every detail.

I propose giving hence the individual liberty to the birds, for group flying has no favourable action in the developement of finding the cardinal points. Only intelligent birds take the direction of their house and the others are satisfied with following these leaders.

It does not seem to amount too much, but by a long series of years of experience as breeder, I know, that the breeder who will follow my advice and give compulsory flight to the whole colony, adopting the 21.860 yds. limit as herebefore proposed, will reap success.

He will develops the recording powers which will improve steadily as the distances increase.

(1 km. = 1093 yds.)

The young bird is not formed the year of its birth, and when considering that the young cocks must fly races the following years, the maximum distance for these pigeons, should be 150 Km.

Fine hens can be entered up to 300 Km. and will be kept then by good breeders for reproduction purposes, and they are often the basis of the finest colonies.

Commonly yearlings females give better results than cocks ; but cocks are better developed and heavier than the hens ; which are more precocious.

Hens are less smart and cannot withstand as long an effort.

The breeder who disposes of a great number of young pigeons, and who does not mind sacrifying some of them in order to get prizes, will do well, to let pair, and even procreate the birds intented for numerous races.

They will have thus an incentive to get home quickly, they will better find their way and will even get lost less easily.

Do not forget however, that any paired young bird knows no rest, eats very little, what has a nefacious influence on its developement.

The paired young cock wastes his strength, and the hen, laying eggs during the training season exhausts herself.

Therefore it is preferable not to pair the best birds, and to keep for following years those, which are several time amongst the first.

THE PIGEONS OF ONE YEAR

Training for these birds will begin in the first days of May.-

Those, which have been raced the first year, will, for the new training be brougth the first time at 20 Km. from the pigeon-house, as explained in the preceding chapter, released at 5 minutes interval.

This first flight will, if possible, be effected in the morning, by fine weather.

No food will be given before the start, in order to obtain a quick return to the dove-cot and in order to train them to this for future travels.

This training will be repeated once or twice before the first race.

The backward birds, which could not take part to the races the year of their birth, will be submitted to the same training as the yearlings, and be released at 1, 3, 5, 10, 15 and 20 Km. in company of old birds.

The individual flights will then take place, beginning once more all the stages over again, as explained for the yearlings ; the 6th stage being 20 Km.

The one year pigeons which are effected to the « natural play » or to the half « widowhood », will be paired only when they are trained.

The paired cocks are keener to come back, so that, avoiding to be led astray, the losses are very small.

The hens of one year intented for traveling, will also be affected so the same cot, but, in order to obtain good results in the races, pair them with cocks who do not participate in the same races, the hen being older or of the same age.

I have noticed, that in variable weather or by great wind, the pigeons of the year get home too late or get lost ; (the losses being rather heavy in this category) when paired they are more regular and get home very well.

In my numerous years of experience, I have observed teams trained according to the method or « widowhood » with pigeons of one year. The heavy losses sustained by these breeders have decided me to advocate the pairing of birds not used for travelling.-

It is however advisable for the breeders intending those one year birds for widowhood, later on, to avoid letting them procreate.

They can be paired and brooding may be encouraged even several times during the stages which will not exceed 300 Km.

In this way, they will not get exhausted, and you will have remarckable birds for speed and endurance for the following years.

An observation, the importance of which is not to be underestimated, is to avoid overdriving any bird obtaining good results, by playing it so long that it does not come back any more.

It is better, to choose a few interesting races, and to play your birds only when they are fit.

Do not follow the example of some breeders, who play their pigeons every Sunday and even in the week.

The daily training in the morning and at night, before feeding, for the pigeons of one year, will take place at the same time and in the same way as for the older pigeons.

Only the cocks will take part in it.

The hens will fly in the morning and at night also, but immediately after the cocks.

These will be released withoud flag, i. e. only the entering to the pigeon-house will be closed during the same time as for the cocks.

They will be allowed to get in again, after this time and eventually go back to their nests.

As a rule ; pigeons of one year are not allowed to travel, through exceptionaily good birds are met with in this category.

They will soon show what they are able to do.

After having travelled, they will be kept for reproduction, in order to fill the gaps occuring in any team through bad weather and annual disaster. It is advisable for races over 150 Km. to enter only half the team, i. e. to divide the pigeons into two groups : one of which will be entered for the following race ; this to minimize the losses.

TWO YEARS OLD PIGEONS

Many breeders think, that 2 years old pigeons can be entered in 900 to 1.000 Km. long competitions.

Others believe, that pigeon of one year can be considered as old.

Training of 2 years old pigeons will be begun since beginning of April.

Those who have competed the preceding year, will be brought to 20 Km. from the pigeon-house, as the one year old, and will be released separately.

Two years old pigeons can be played in several ways : « Widowhood, natural, speed and half widowhood ».

They will be allowed to travel from 5 to 6 hundred Km.

They will be entered only when properly trained. Here, the breeder only is responsable.

Any pigeon, not perfectly fit, i. e. which has not all its powers keyed to the utmost, will be tired soon, and will spoil its training for any future competition. It gets back to the pigeon-house down hearted, and will be easily lost.

2 Years old pigeons as many others will take their daily exercise in the morning and at night, and will be trained according to the method used in that particular team, widowhood, natural, half widowhood or speed.

THE WIDOWHOOD

These birds will be paired on the first of march, and will remain with their mate 21 days, though taking their daily exercises.

This is the exact length of time necessary for the hen to lay eggs and to brood them 10 or 12 days as a maximum.

The hens will then be taken away, and the cocks be kept 1 or 2 days longer. The eggs will be taken away as soon as abandoned.

The nest is thus empty ; the cock left alone.

From this day on, quite a special life begins for him ; for the hen being taken away, he is an actual widower.

The hen will not be brought back to the cock during all the travelling season. She will not even be seen or heard.

The hens though being put aside and sequestred during the travelling season, fly very well by this method, and I shall even add, that many times every year, I have registered many victories with such hens.

I will come back to this point later on, when describing the complete widowhood.

NATURAL OR SEMI-WIDOWHOOD

The birds will be paired in this case about 4 or 5 before the first competition.

They will have their flight alternatively in the morning and at night. The cocks will be released at first, using the flag as explained, in order to keep them in the air during one hour.

The hens will then be released without using any flag, but the entrance of the pigeon-house being closed.

The cocks, trained at half widowhood, will be separated from their hens, as soon as they have brought up the first youngs, and about 10 days after brooding the second couple of eggs.

In the first stages, the pigeons will travel without seeing the hen neither before putting in the hampers, nor before their return when training.

In view of the first competition, the system will be changed.

By this, we mean, that the last Sunday before the putting in the hampers for prizes, the cock will be given its hen back, and on the day of sending the pigeons off, he will cover the prescribed 20 Km., being put by its hen when getting back.

This short flight is important, and will be carried out weekly by all birds brought up in half widowhood, and having to participate in the competitions.

It will be seen, that hens of these pigeons, through, being separated in good hygienical conditions during the racing season, obtain good results in competitions with this method.

Care will be taken however, that these do not pair amongst themse'ves, and contrarily to the method of complete widowhood, they may be heard by the cocks.

A simple method of avoiding any burst of tenderness between hens, is, to built outside the pigeon-house as many large cells, as you have widow-pigeons.

If for any unexpected reason the cock cannot take part in a competition, and that the hen has already covered the 25 Km. hereabove mentioned on the usual flightline, let her cover this distance separately, so that she comes back to her mate when getting home.

This method is a very successful one and can be practised in a simular way for the hen.

THE SPEED DOVE-COT

Pigeons intented for the speed races will be paired about the I Februar and will have a daily exercise in the morning and at night, as soon as the short distance contests are near.

This exercise will be effected in the same manner, as for the « natural » and « half widowhood » method ; cocks first, then hens.

Birds which, as yearlings or one yearers have already travelled, will be brought 20 Km. away and released separately.

This selection is composed more of males than of hens, for birds intented for speed contests are often 2 cocks for a hen ; they will not be released together.

In this method, all the systems can be used and training will take place as explained to the chapter specially reserved to that system.

OLD PIGEONS

Old pigeon which cannot be considered yet as « Old Timers » will be trained according to the travels they will have to effect.

Is considered as old pigeon, any bird 3 years old or more.

They will be allowed to compete in any competition even up to 900 or 1.000 Km. taking into consideration, the distances covered the previous years.

On these pigeons 3,4 even 5 years old, is often built the renown of a team.

It is up to them to balance the budget and even to leave a benefit.

We consider as « OLD TIMER » pigeons 4 or 5 years old, having covered all the distances in any weather.

They will not be looked after as yearlings or one year-ers.

These birds will partake yearly in 3 or 4 long travels and their preparatory stage will amount at least to 200 Km.

They are slower than young birds, but are considerably more efficient in bad weather.

They dispose of experience of stayes covered in any weather, by any winds, of nights spent in the open.

They do not get downhearted if they are fit to withstand an effort, lasting hours and even a whole day.

When such birds are entered for some long race, they must be carefully prepared and put in the hamper in the position they prefer and which you will have observed previously.

In order to keep them fit, do not pair them too soon, to play them natural or in half windowhood, and make them not feed them as in the beginning of the season, so as to have 8 or 9 quills for the great contests, and to have them so in the best condition.

GENERALITIES

Pigeons regulary trained will get no food before being released. This rule must be strictly enforced, for if the birds had food in the crop, they could not fly fast, and would not improve by the exercise.

Those which must participate to a contest, will even instead of getting a supplementary ration on the day of putting in the hampers, be fed for the last time in the morning.

Many breeders do the contrary, but they feed birds too much. This is absolutely wrong.

Very often, birds which have received a double ration reject during the transport from the pigeon-house to the place from where they are sent off, all the food they have absorbed.

The colonel A'Speculo, the président of the « Jeune Hirondelle of Liége », interviewed once about this point by a novice, answered : « So much the better, old chap ! Your pigeon is rid of its superfluous ballast. »

The colonel was right, for this habit of overfeeding the birds before the start can bring about serious diseases, such as heart trouble.

Train only pigeons which are fit ; keep only a small number of birds ; make them familiar, feed them regulary, rationnaly in order to avoid having too fat birds. Take note of weather, the position, the stage covered, the day your pigeons have been successful and specially the weight, for this is very important for your future observations. The weight must indeed be in proportion to the wing area. This will be explained later on.

When the birds get back from their daily practice, do never scare them ; always talk with them, do not speak too loud, and if you can spare a few minutes, take them in your hands, morning and night, considering them as friends and never being hard.

A daily training at about 8 o'clock in the morning for about an hour is excellent.

You will so train your birds to morning flights.

Never train irregular or abnormal birds ; you would lose money and time.

Fitness

Perfect fitness can be obtained in a pigeon as in any other being only by careful training.

This readiness is only momentary and reaches its maximum thanks to good food and besides to fidelity to the pigeon-house.

This readiness, this attractive and affective power can be obtained only with perfectly healthy pigeons, free from any defect, and the moulting of which has been easy.

If these conditions do not exist, readiness can not be obtained, and if it happens in a bird, it will only be for a short time.

Never can an unhealthy bird concentrate in its organism, the energy required to cover a great distance.

The readiness is not latent, and the best bird cannot reach it if not properly fed and taught to be faithful to its pigeon-house.

HOW TO DISTINGUISH A PIGEON THAT IS FIT

The readiness is not always easy to discover, and one must be a keen observer to notice it in a bird.

In order to observe it, you must know its usual condition, in order to be able to compare and so to a certain any more or less favourable change, in the possibilities of the bird.

When I told you to hold your pigeons in your hand every morning an night, it was not only to make them familiar, but also to know them perfectly well and to get well acquainted with their physical built, until you are able, all your birds being on the floor, to put each of them back in its cell, without the help of any light, i. e. as a blind man.

Repeat this at various times of the year, observe the various organs in the morning and at night in order to be able to judge a certain change taking place in the bird, and having an influence on the potential energy.

You will take note of every observation in order to make logical deductions, which will prove precious when the travelling season has begun.

WHEN THE PIGEON PECULIARITIES IS FIT, IT SHOWS THE FOLLOWING PECULIARITIES : THE PLUMAGE

The feathers are very compact and covered more than in the nomhmal state, with a sort of grey mealy dust, extending specially to the superior mantle and to the secondary quills. It can easily be removed.

Take the bird in your hands, open the wing, which must be full, i. e. which must show no opening between the quills.

You must then observe a small channel on 2/3 of the length of the second feather, this being formed by the bristles on the inside.

When the first feather falls, the bird must be raced, and it is advisable to wait, that said feather has grown again to at least 2/3 of its normal length.

This first feather is very dangerous, for sometimes it brings about a great change in the readiness ; this having an influence on the whole organism. Having observed it a thousand of times, I shall say that 95 % of the birds cannot get good results as long as the first feather is not entirely grown back to 1/3 of its length.

Considering the fact that, when the fifth feather falls, the season must be over, you can frankly race a bird, having the hereafter qualities, if, from the first feather to the fifth, the fall takes place on the day of sending the pigeons off, and if the bird is to be entered in a speed contest.

By this, we mean a contest necessitating as a maximum 2 days spent in a basket.

Should the feather fall out one or several days before the start of the pigeons, the bird wil not be entered, and one will wait until at least the following week for entering it.

This period is the necessary delay to allow the feather to grow back to 1/3 of its length.

The tail will be extraordinarily clean and very narrow, so as to let only one feather visible. Open the tail and see that it does not show any empty space. These spaces are calle « cobweb » in Belgium.

Pass the hand under the wings, and in no case you must find « down that is yet to come ».

Both wings must work easily.

The same, applies to the down around the rump. The button of the rump will be dry and rosy coloured.

The down covering as a rule, this rump all the year, will be developed, white as snow, without buds, and leave the impression, the hand being introduced in it, to be wet.

All the feathers must be compact, silky, close to the body, shiny and form straight lines with rounded outlines.

It happens however that a bird though having all this qualities does not obtain good results in racing; the primary quills must then be closely watched and they must reveal no parasites, as explained in the first lesson. The microscope is of great use in revealing these parasites.

I am the first to have observed them and have used erroneously the word microbe to describe them, in my first lesson.

In a pigeon which is fit, the head seems smaller and the feathers seem to be denser.

This head is really fine, and gives the impression that the bird is washed, combed until the feathers seem to adhere closely to it.

If you are not sure yet, that the pigeon is fit, put it in a basket, by preference in a skeleton case. Leave it in a calm place, far from the pigeon-house, so as to avoid any incident in the house, troubling the bird ; and then, after half an hour, have a look at the bird and see if no feather is ruffled.

The ears will not be discernable. A similar inspection wil be held before each race.

When the pigeon is full of fire it is liable to variola and thrush, and if the bird begins this disease at the ear when sent off, without the trouble having been noticed, you will surely lose not only your entering money, but also the bird.

See for this disease the second lesson, about the power of finding the cardinal points.

BEAK

The beak will be smooth and clean, glossy shiny on all the length and all the upper part, with a feever spot at the extremity.

The nasal caruncles and the eyelids will have a white pigment with small rosy points.

The nasal chambers wi'l be quite free and well open ; the slightest disease can prevent a bird being classed.

Therefore it is useful to inspect them at the same time as the ears.

The mouth split ; inside the bill will be well opened on its whole length pink. This shows that the breathing organs are fine and sound. The diseases of these organs can hamper the bird and prevent it following its competitors.

RESPIRATORY CHANNELS

The palatial split or mouth split must open and close constantly ; and show a pink colour from the dupplicature to the first part of the beak.

To ascertain that the channels are not obstructed, open the beak which must be pink in the inside in its first part, and free from any yellow or white spots.

This palatial split, serving with the nasal chambers to filter the inhaled air, before it reaches the lungs, must have all your attention.

The air contained in the air sacks and the lungs, passing through the nasal chambers, the palatial split and the larynx, gets warmer and avoid these organs getting cold.

Fig. 32.

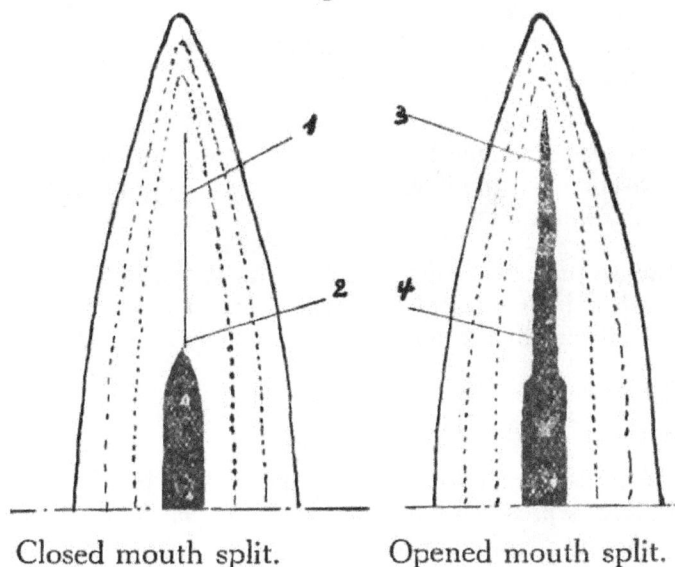

Closed mouth split.   Opened mouth split.

When the mouth split is obstructed, the bird is compelled to open the beak constantly and the respiration is not normal.

The bird does not get enough oxygen to purify the blood reaching the lungs, the impurities gather there and this causes a general uneasiness, making any long flight impossible.

Each time respiration troubles occur, tha palatial split closes and is covered with white or yellowish spots.

The bird seems, when fit, broader and rounder than in the normal state.

When in the hand, you can feel the air pouches greatly swollen and feeling slightly with the fingers, you observe that the bird is harder though lighter at both sides of the breast-bone, at the front part about 2 centimetres from the end you will feel as two springs.

MUSCLES

Both pectoral muscles rest in the angle formed by the breast-bone and the inferior face of the sternal plate. They are very interesting.

I told you in the first lesson that they rested in this places tormed on both sides of the breast-bone. Their contractibility is affected by the movements of the bird and by the food.

These muscles develop and get strong according to their degree of contractibility.

Once the maximum of contractibility has been reached, the muscles press on the flesh, especially on those covering the plate of the sternum, along the breast-bone. This flesh is pressed back, and forms 2 little pads which will remain there as long as the organism is full of strength and that no alteration will be introduced in the birds life (see picture 33).

These little pads are easy to recognize and when you hold the birds in the hand, they give the impression of holding a well inflated ball.

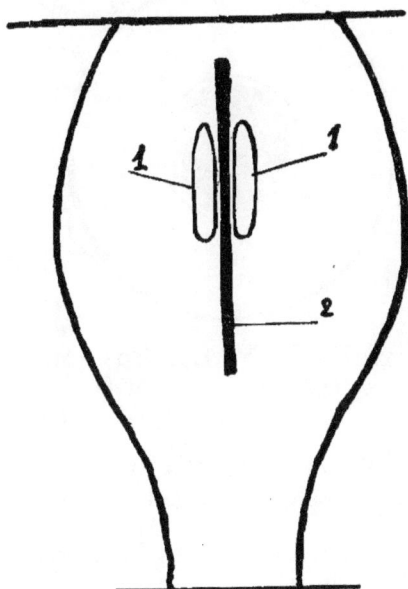

Fig. 33. — Pads of a fit pigeon.

The flesh on the whole body will be free from scales, hard and pink.

THE EYES

The eyes must be carefully studied.

This will happen as many times as possible in the normal state, in order to observe the augmentation in the number of granules, composing the various circles of the eye.

The eye reflects the qualities and defects, the diseases and the health of the birds.

It is composed of numerous pigmentary coloured cells-which form the circles.

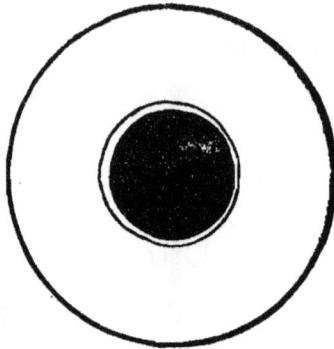

Fig. 34. — Normal Position.

Picture n° 34 shows a pigeon's eye in a normal state, i. e. when the bird rests and is not specially trained.

The eyeball is nearly motionless in its real size and is in the centre of the eye.

The colour is regular and the circles surrounding it, remain in the same conformation and will remain so as long as the optical nerve will not exerce an influence on them.

As previously told, the optical nerve transmits to the brain the impressions perceived.

When the bird receives a tonic food the nervous system brings about a transformation in the eye as follows.

The pigmentary cells under the encreasing nerve pressure of the nervous system, and quite specially of the comb, swelling under the rush of blood produce in the eye different pressures, change its aspect, and concentrate near the outside edge, which takes a deeper hue.

As shown by picture 35, the eyeball has decreased in size, the circle of correlation has encreased, the circle

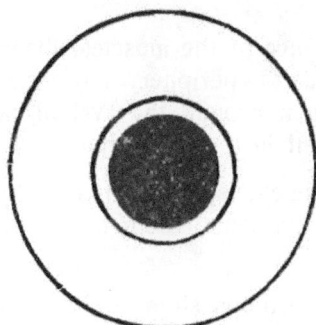

Fig. 35. — First apparition of getting fit.

of the iris is more coloured and the fifth circle is more visible.

Picture 36 shows the eye of a pigeon fit for travelling.

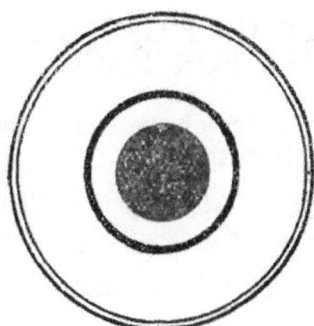

Fig. 36. — Searched position.

I have mentioned, that the comb has the property to swell under the rush of the blood, what produces inside the eye variable pressures, and gives the organ an other appearance.

The optical nerve serves as a conductor to the sensibility and transmits moral impressions.

It works on the eyeball and gives it quick alternative movements, contracting and dilating it, even diminishing it to its utmost.

Under the influence of the muscles, the pigmentary cells are rejected towards the periphery and disclose a new circle, without any pigment around the eyeball, which keeps the primitive colour, but in a deeper shade.

The outline of the eye is sharply marked and a fifth circle appears, often of the same colour as the correlation circle, or in a deeper shade.

These various positions show all the circles well determined : eyeball very reduced, makes a sort of star ; the correlation circle having a great area ; the iris circle, having a deeper shade ; the fifth circle clearly visible.

The eye of the bird having these qualities is dry, lively, has smooth eyelids, showing little pink spots, the lower part of which is duller.

DETAILS OF PIGEONS BEING FIT FOR RACES

In fit birds, the wrist of the wing, sometimes called shoulder, separates from the body.

The feet are light red without any pellicle, the nails and fingers will be smooth and clean : they feel very hot.

The excrements must be hard and excreted in small balls 3/4 grey and 1/4 white.

Turn the pigeon on its back, and examine the breast-bone, blowing for this on the feathers.

This breast-bone must be clean, free from pellicules, and if there are any, they must come off easily when rubbing with the finger.

You will also observe the two little pads, before the breast-bone, one on each side. It gives the impression that the body of the bird is just like a small well inflated ball. (See picture 33).

The little blood drop immediately under the skin along the breast-bone is present in any bird, either at rest or in readiness, i. e. having received a special food preparing it for travelling.

This blood drop shows however the progress in the training.

1) When it is at the rear of the breast-bone and big as a hemp seed, it shows that the bird is not fit for a fast flight.

2) When it reaches the center, and is smaller and moves continually, it is a sign that the bird is fitter, and that it soon will be ready.

3) When it reaches the front part, as near as possible near the point of the breast-bone, and is small like a rape-seed and moves constantly, it shows, that the bird is in a perfect condition, in as much as it fulfills all the other requirements as to its physical conformation.

4) When it gets bigger, remaining however in the front part of the breast-bone, it shows that the bird gets less fit ; though being still in good condition.

5) When divided in 2/3 spots, moving to the back, it shows that the bird is no longer fit.

Never enter a bird in a declining form, but always in an ascending form, as shown by the mysterious point described to hereabove.

These observations can be controled by the thermometre indicator of condition, if its accuracy is perfect.

We can supply such apparatus, guaranteed accurate.

APTITUDE TO THE FLIGHT

There are many ways of observing the physical dispositions of the birds.

1) VISUAL EXAMINATION

A few hours before the start, act in the following way.

Take a few exhibition baskets or skeleton cases (small wire rods of 5/6 mm disposed 5 cm., from each other), in order to avoid the birds damaging their wings and enabling you to observe the birds without disturbing them.

All the birds intented for a competition will be put each in such a basket, during at least one hour, in a quite place, far from the pigeon-house.

Do not disturbe the birds ; then observe them without taking them in the hands.

The fittest bird will be the one having glossy feathers, none of these being ruffled.

The bird will be clean, well combed, with however some dust on the back, between the shoulders.

With some experience you will soon notice the best conditioned bird. It will be proud when you come near it, the two wings will be placed in an horizontale line, without that the extremities cross themselves ; the tail will be narrow, and will not be larger than one feather.

The eyes be bright, and if you come near the basket, the pigeon in condition will not be afraid.

It will watch you, and always look the way you stand.

The beak will be clean, specially the extremity which will show a fire point.

The caruncles and eyelids are pigmented with small pink spots.

The feces will also be carefully observed. They must be hard, expelled in little balls as big as a horse-bean, 3/4 grey and 1/4 white.

If the feces are more or less liquid, the bird will not be entered and be kept so until the excrements are satisfactory.

2°) GOOD MARKS

When the whole team is in the air, being however free to come or go according to the wishes, of the birds, a cock shows good disposition, when he comes to the landing board, looking briskly to the left and to the right and if, when seeing an other bird even very high, he takes then the air once more, clapping his wings, climbs higher than the other bird, as a cock chasing his hen about to lay eggs, claps its wings constantly. It will then come back to the house ,and cruise over it, wings spread out as a rule.

After a few circuits, he comes down, dives with closed wings until he reaches the landing board, and makes a lightning entrance.

After a few minutes spent in the house, examining what happens there, specially around his cell, he goes back to the board, and makes once more the same movements as here above described.

These movements will be repeated as long as the bird is in condition or in readiness as we call it.

3°) REFERENCES OF DISPOSITION

For the speed contest, the bird will be brought to a hundred metres (meters) from its house, and will be released alone. You will observe if there is no other bird aloft in the neighbourhood, what would, perhaps, spoil your experience.

The purpose is to show how keen the bird is to get home, in good conditions.

Will be in good conditions, any bird flying away rapidly without any hesitation direct to the pigeon-house, keeping a medium height.

When the bird is released by contrary wind, it can make a circuit; but this one must be of a small radius, without that the bird be driven away several hundred metres in the opposite direction. As soon as it has required the height and is on its way back, try to observe carefully your pigeon. If it meets any other pigeon it must neither let himself be led astray, nor prolong its flight.

Far from accompanying the other bird and even from going its way, if your bird possesses all the required qualities, it will remain alone and keep flying rapidly.

Once in the neighbourhood of the pigeon-house, it must dive without any hesitation, never even stopping at the landing and get immediately inside.

4°) INDEX OF OBSERVATION

There is an other way of making sure that a bird is fit.

You must come very noiselessly to the door of the pigeon-house, in order to be neither seen nor heard by the birds.

As we know, the door must be provided with a blue pane. This will allow you to watch the birds without being seen. If you practise the « Great Widowhood » full separation or the « Half Widowhood » when you notice a bird which occupies the nest cage, cooing meanwhile, going and taking suddenly possession of a neighbouring nest, this is a sure sign that the bird is quite fit and that it has energy and power enough.

Such a bird is sure to get prizes if it possess the previously described physical qualities. But if you practise the «natural» method, if you observe your bird lying beside its cage, with half open wings; sometimes also with the wings far from the body, or lying on one wing while it looks around the whole pigeon-house, ready to dispute its place to any intruder, this shows also it to be fit.

5°) READINESS AS SHOWN BY THE BATH

In order to obtain regulary prizes, the pigeons must be bathed.

They will have their bath, (the whole team) on the day folowing the races. You will be able to observe a fit bird at a glance.

The birds will be held during at least half a minute one by one, in a pail 3/4 full of lukewarm water.

They will be immersed till to the eyes.-

Take then the bird out of the water; do not dry it or open its wings.

Any bird fulfilling all the other qualities, will be fit if it does not retain any drop of water on its feathers.

6°) DISPOSITIONS

When the season is rainy, if you want to judge the condition of one of your birds, let it fly, and keep it aloft for about 10 minutes, using therefore the flag or any other contrivance.

When the bird enters the house, you will recognize immediately a fit bird at its dry feathers.

These birds only will be entered for racing.

The others will be put in observation for the future moulting.

CONCLUSIONS

These hereabove given informations, are of great importance and constitute 6 different ways of discerning a bird thal is in a good condition.

Any fancier, before entering his birds for races, will judge if they possess all the required qualities.

Much money will be saved by following these hints.

If you use my advice, you will avoid many defeats and will on the contrary reap any amount of success.

Classification Following my Method

Now we must classify the birds and I take this opportunity to remind you of the qualities a pigeon must possess
and which have been explained in the first lesson. (Ideal
conformation, page 73-85).

The pigeons are classified as follows :

Excellent in any way :

1°) Class reserved for birds showing qualities as good
reproductor and good traveller (able to be successful from
short distances up to 1.000 km.).

2°) Good reproductors and birds successful from 300 up
to 900 or 1.000 km., will be classified in the same category.

3°) Other good reproductors, successful from the first
stage up to 300 km. will be registered in the same category
as excellent in any way.

4°) Good reproductors, but not good travellers owing to
accidents etc., will be classified *for reproduction only.*

Birds which are not good as reproductors but good travellers will be classified first class according to their aptitudes :

A) 1ˢᵗ category : all distances.

B) 2ᵈ category : long or half long distances.

C) 3ᵈ category : speed.

Will be classified as second class : excellent birds having every qualities, but unfit for certain reasons such as : overdriving, weakness, bad moulting. These reasons bring them momentarilly in this category ; but there is hope to get them fit again for the following season.

They will be specially looked after.

We shall have also an « irregular » class, including birds seldom successful, and difficult to get in into condition. They are unapt for travelling or reproduction.

EXCELLENT IN ANY WAY

Will be classified in this category; any bird having the required qualities as traveller or reproductor.

These qualities are :

State of health must be perfect.

It is shown by the look of the feathers, and by an easy moulting.

Holding the pigeon in the hand, open the beak, and examine the mouth split. This one must be free, pink, free of any white or yellow spots, opening and closeling at will.

When opening the feathers on both sides of the breastbone and at the wings, the skin must show pink.

A good, well balanced pigeon, put in a basket, leans slightly forward. It measures the same length from the point of the breast-bone to the extremity of the tail, as from the under part of the feet to the top of the head.

The fore-arms are equally thick and bent so as to form only one line when looking at them under any angle.

It is easy to handle it, and will balance on one of your fingers.

If the bird is a cock and not affected by tickling, it will slightly strech the feet backward when balancing.

His feathers will be glossy, bright, velvetty. The bird will not be fleshy and gives the impression of having only bones and feathers.

The primary and secundary quills will not be undulated or split, and will show no « cobweb » or holes.

The *down* on the rump, on the sternum and under the wings will be white and full blown.

Let us now consider the wings :

One of the main qualities of a good pigeon is, having the elbow joint close to the body. This joint must be strong, firm and supple.

If they are close to the body, they indicate a bird able to win in any weather.

The closed wing will be broad from the elbow joint to the wrist, will have a sufficient thickness, and when bent downward when open, you will feel a certain resistance.

The primary quill feathers are very broad rounded for the male, more pointed if the bird is a female.

The eight, ninth or tenth are equally long and superposed, so as to build only one feather.

The stems (shafts) will be fine and flexible; the part entering into the hand of the wings will be quite white.

They will be neither split nor undulated, and quite free from parasites. These ones will sometimes stick by hundreds to the shaft.

The mantel will be very thick and the secundary quill feathers large and not too long, in order to have a form similar to the wing of the swallow.

A perfect pigeon will have a very thick and resistant breast-bone, varying between 7 tot 9 cm. length from the fore point to the fork.

This bone will be in direct line with the point of the front of the wrist, and will show a slight curve and a slight sharp end towards the crop.

It must be neither too curved nor too flat.

The longer and thicker it is, the fitter the bird is for long distance travels.

A well built breast-bone begins as a sharp and slightly curved bone and ends at the fork, to which it seems to be united.

For the cock, the breast-bone will show no break from the bone to the fork.

A bird with a breast-bone bent to the right or to the left, even forming an ⚬⚬ , does not « ipso facto » lack, qualities, but it must not show any deep cavity, spoiling the perfect form of the bird.

For the male, *the fork* will be very short and well united at its extremity; both bones will be straight and of equal length.

When a bird has an arched fork, or a fork with one side shorter than the other, it is just as well to disclassify it; it will never give good results, being unbalanced.

For the female, the fork can be 2 cm. long, and show a break, when leaving the breast-bone.

A female with all the required qualities, but having an arched fork, will be reserved for reproduction, but she will not be allowed in competitions.

Independently from the sex, *the breast* must be broad and rounded, showing that the muscles are well built, and that the air pouches are well developped.

For a male, when passing the fingers under the fore-arm of the wing, you will notice a little pad as big as an almond.

For a hen, the same pad exists, but smaller.

The perfect pigeon, in the hand, gives the impression, that one holds only the tips of the wings, the feet and th tail.

All the weight is forward, the flesh is hard on both sides of the breast-bone.

A well built *body* must be ovoidal, well rounded, and diminishes in sise, pear shaped, at the rear end.

The back must be broad and round. A pigeon in good condition put in a basket shows a slight distension on the back.

A flat backed bird must be discarded as this retains the rain, and will never give good results.

The neck must be short, thick and as strong as possible; this part being subjected to great stress after long hours of flight.

When the bird is aloft, the head is moving constantly.

The tail is the rudder of the pigeon, and as the primary quills, they must show no flaw.

The shafts and bristles will be carefully examined, and will have the same peculiarities as the wings.

A pigeon with a long tail is able to be successful by South wind. A bird with a short tail will reap success in North wind.

A bird, having a middling-sized tail will succeed in any weather and by any wind.

The *head* will be midding-sized, slightly flat on the top, and ends in a graceful curve toward the back.

The *ears* concealed as much as possible.

The *eye* will be well set in its socket. It must be bright, shining, placed high in the head, as near as possible to the skull.

The look will be slightly forward and downward.

The eyeball will be above the line prolonging the beak's split.

The *eye* must be perfect i. e. possess a complete circle of adaptation, this one being as thin as possible.

The circles will be well distinct. They will never neither encrouch one on the other, nor be pigmented.

The whole eye seems always larger in a fine bird. Its gaze is always directed towards to you.

The eyelids will be either grey or white, sometimes pink, but this is seldom; they will enclose as much as possible all the eye.

An intelligent pigeon has a broad forehead, very high with a small protuberance of the upper part. The forehead declines gradually to the morels without making any fold.

In a good bird, the beak is of an average size. Both mandibles cover perfectly each other, without leaving any hole. The beak is provided with middling-sized caruncles. These must not be cracked, this impairing the power of finding the cardinal points.

By old pigeons, they must still be smooth though bigger.

When a bird has a superior mandible larger then the other one, it will be disclassed, owing to the fact that it cannot pick up easily the seeds; this producing a general weakness.

A long beak is a sign of a bird fit for covering long distances, while a short beak shows a bird apt for short distances.

A middling-sized beak denotes a bird able to succeed on every distances. But be careful.

Feet will be middeling-sized, with good muscles at the thigh, getting gradually thin towards the knee.

A good bird must possess well built legs. If it gets down during its flight, it is owing to its feet being tired, or to a great thirst in a flight, under a burning sun.

The air-pouches give a rounded shape to the chest.

You can feel them easily on the front part of the breast and on the back of the spine.

They must be well inflated, for it is thanks to them and to good shaped wings, that the bird is able to remain aloft and to cover long distances without being too much tired.

As a rule, when a pigeon possess one of the hereabove stated qualities, it possesses also several of them if not all.

First Class. — TRAVELLING PIGEONS

Birds having no qualities as reproductor, can as well as those classified for general excellence, be divided in 3 categories A, B, C.

We shall classify in these categories, any bird possessing the qualities making a good traveller, but not having those of a reproductor, owing to one of the following defects.

a) Uncomplete circle of adaptation. A 1/4, 1/2 or 3/4 of this is left only ;

b) The eye is deprived of the circle of adaptation.

A. 1ˢᵗ Category. — GOOD FOR ALL DISTANCES

The bird must possess the following qualities.

Be *healthy*, have a rounded *body*, glossy *feathers*, velvetty to the touch, a narrow middling sized *tail*.

Well developped *chest*.

In the open *wing*, the primary quills will be in a line with the secondary ones, the three last equally long superposed as to show only one feather.

The breast-bone, very strong, minimum 80 m/m. long for the male and 70 m/m for the female.

Fork of an average size, very strong, straight and narrow by the cock.

It can be 2 cm. long for the hen.

Head, middling sized, slightly flattened at the upper part.

Forehead, broad and high, slightly swollen at the upper part.

Beak, middling sized, the mandibles must show no opening.

Caruncles, clean.

Ears almost unvisible.

Neck, strong and thick.

Corpulence; medium.

Eye; see picture 28, 29; or with many colours, well set in their socket, medium-sized eyeball, correlation circle well marked, circle of the iris medium-sized, richly coloured, 5th circle visible, fine eyelids, completely enclosing the eye.

B. 2ᵈ Category. — GOOD TRAVELLER FOR LONG AND SEMI-LONG DISTANCES.

These pigeons must be healthy; with a round body, glossy or velvetty feathers, a narrow tail of an average length.

Chest; well developped.

In the outspread *wing*, the primary quills will be in line with the secondary ones. The three last are equally long, superposed as to show one feather only.

Breast-bone ; strong and 85 ᵐ/ᵐ long for a cock, 75 ᵐ/ᵐ long for a hen.

Fork, medium sized, very strong, straight and closed by the cock, and can be 2 cms. long by a hen.

Head strong, slightly convex towards the rear, getting thinner towards the neck.

Forehead; broad and high, with a slight protuberance on its upper part.

Caruncles; clean.

Ears ; practically concealed.

Neck; strong and thick set.

Appearance; rather corpulent.

Eye; pictures 28 and 29 or many coloured, well set in its socket, eyeball in the small side; circle of correlation finely marked ; iris well developed, rich in colours ; 5 th circle not very visible; eyelids seeming more oval.

C. 3ᵈ Category. — GOOD TRAVELLER FOR SPEED CONTESTS

Must be healthy, have a good round *body,* glossy velvety *feathers ; a narrow ; tail* medin-sized.

Chest well developped.

Wings; are springy; when outspread, the primary quills are in line with the secondary ones; the three last of equal length, superposed so as to look like one feather.

Breast-bone; very strong, 75 $^m/^m$ long for a cock, 70 $^m/^m$ for a hen.

Head; small large forehead, high and convex at the upper-part; short mandibles leaving no opening; caruncles clean, ears concealed.

Neck; strong and thick set.

Appearance; smaller.

Eye; like described in the pictures 28 and 29 or many coloured; well set in its socket; eyeball well developped; correlation circle strongly marked; circle of the iris thin and richly coloured; 5th circle well marked; eyelids rather developped.

Second Class

The bird having all the qualities required for birds susceptible to be well classified but rejected owing to faults such as : overdriving, rough feathers, bad moulting, split quills or quills having openings, weakened muscles, black tongue and anaemia will be classified in this category.

IRREGULAR PIGEON

Any *unhealthy* or *badly balanced* body or *badly conformed bird* will be classified as irregular.

Body badly rounded, too narrow chest, joints fitting badly, flat back, short breast-bone, too flat or too deep, or presenting a deflection weakening the sternum, or forming a too marked break between the sternum and the fork.

But do not mistake a ∽ formed breast-bone, with the hereabove mentioned defects. This one is not serious in a travelling bird, and does not affect its balance.

Rough feathers; dry; bad moulted; split quills, like a pencil or like a cobweb: too big shafts, and yellow down mark these pigeons. Unequal or bent *fork.*

Wings; too short, too thin either badly shaped or badly jointed or badly curved.

Forehead; too narrow or with a marked break.

Caruncles; too big or cracked.

Beak; hollow or with openings between the mandibles.

Neck; too long or too thin.

Eye; badly set, badly formed, too much in front, too mush at the back, too fleshy or too wet, circles intermixing; iris too pigmented.

HOW TO RECOGNIZE A GOOD PIGEON

You will now be able to classify a bird in the required category.

It is not always necessary to take the bird in the hands; observing it will be often enough to have a good and sound opinion.

This will be done :

1) In the hereabove described manner.

2) By looking at the bird in a basket.

3) In the pigeon-house.

4) In the bath.

5) When in flight.

6) When it is raining.

LOOKING AT THE BIRD

When you want to judge a bird in its basket, the feathers will give you a sure indication as to its condition.

The good pigeon has always glossy, rich, velvety feathers.

The bird is well balanced, the body leaning slightly forwards, the eye will be bright, full of live, well set, with complete circles, these ones being distinct; the head will be of an average size, the caruncles being clean and smooth. The forehead will be high and broad, with a slight protuberance on the top; this showing the intelligence of the bird.

The neck will be thick and strong. The chest will be round and broad. The full wings will make an horizontal line and will not cross at the extremities. The tail will be narrow and medium-sized. The feet will be strong and not too high.

IN THE PIGEON-HOUSE

When you come near a good pigeon, it remains calm and familiar. When you want to take it, if you make it understand your intention, it will allow itself to be picked up.

It remains in its cell, even when the whole team is free, for i tseldom flies, out of the training hours. It likes its house and doesn't visit the neighbouring cells.

If it leaves its cell, it will not fly. It will run away, even for the weakest bird. One could believe it to be conscious of its own value.

During the rest-time, when the food is given to the whole team in a crib or on the floor, it will eat slowly, picking up the seeds which are far from the lot of birds ; these ones pick the seeds glutonly up.

A good pigeon is a good breeder, and will bring its youngs carefully up. It will look after them and be very careful. Sometimes, it will even feed the youngs of its companions and will never chase them.

IN THE BATH

When the moulting is over, and the whole team takes a bath, dip them one by one up to the eyes in a pail of water. Leave them without shaking for half a minute. The dryest bird is the best. You will not notice a drop of water on its feathers.

IN FLIGHT

When the whole team is aloft, the good bird will leave the house first and lead the team during the whole flight.

The position of such a pigeon is different from that of the other birds. It will keep leaning forwards and downwards as if ready to come down any moment.

DURING THE RAIN

When the moulting is over, release all the birds in a normal, but not too cold rain, and keep them aloft for 10 minutes.

After this, let the birds come down and examine them one by one.

The good pigeon will not show a drop of water on its feathers, mainly on the back and on the front part of the crop.

An irregular bird with either rough feathers, or having badly moulted, or with a flat back will be wet through.

This shows that this bird is unfit for competition-work.

———

These 6 ways of judging a bird are the result of 40 years of practice. I do not wish to take them along with me in the next world, but want all my breeder-friends to share them with me.

When you select the birds and pair them, be severe. Do not keep any useless bird, and let procreate only birds worth being taken in consideration under every condition.

This advise will make you successful and will encrease your fame as a breeder.

———

Diseases and Accidents

The pigeon is as liable to diseases as any other being. It must be well looked after. It needs much air and light. Each bird uses about 1.20 l. oxygen per hour, or 30 Hls. a day.

A practical way of judging if your pigeon-house is suitable, is, to enter it in the morning, before releasing the birds. You will thus be able to feel the temperature, and to judge if the air is all what it must be.

No smell must be noticed, viciated air is fatal to the birds. It slows down the organic combustions and intoxicates the birds.

The house must be neither hot in Summer, nor cold in Winter ; these differences in temperature being very detrimental to the health of the birds. A constant temperature must be maintained.

Great difference in temperature will bring about serious troubles.

Cold provokes a deperdition of calories, which can be compensated by a supplement of food only.

Heat congests the skin and anaemiates the inside organs.

Air is necessary, but avoid carrefully any draft and dampness. These two factors decrease the value of the birds in considerable proportions.

The situation of the pigeon-house is of the utmost importance : we shall try to orientate it to te the South or South-West. This position admits the first sunrays.

The diet is very important also.

Give all your attention to the manner in which food is given.

It's through and with the food, that a team often gets contaminate.d It is advisable to feed in a small trough placed in the cell, during all the racing season. The trough will be taken away and cleaned after each meal.

The food id given only in special dustproof troughs, and well preserved from any contact with the faeces.

Drinwater can also become a cause of contamination. It will be changed at least once a day. The trough will be cleaned once a day at least.

Notwithstanding all your cares, your birds can be contaminated by other birds neglected by their breeders ; this during the travel-season.

When getting home, they will be carrefully examined and get a bath on the following day.

When you observe the outbreak of a diesase in one of your birds, separate it at once from the rest of the team. Doing this you will prevent all the team getting contaminated, and allow the diseased bird to be the best looked after.

Several diseases can attack the pigeons ; some are contagious, some are hereditary, other are sporadic.

The most frequent ones interest the digestive and respiratory organs.

DESCRIPTION AND TREATMENT

I will describe the most frequent diseases. You will find in the hereafter given notes the symptomes of the principal diseases and the cure of them.

The subject is divided into five chapters.

1) Diseases of the digestive organs.

2) Diseases of the respiratory organs.

3) Diseases of the organs of the circulation of the blood.

4) Various diseases.

5) Parasites.

1°) DISEASES OF THE DIGESTIVE ORGANS BLACK OR BLUE TONGUE

This can be the effect of the natural conformation of the bird or of a disease.

In the first case. no attention will be paid to it ; the general condition of the bird showing the state of health.

When it is the result of a disease, the mucous membranes of the beak are affected by it ; they are black or blue ; the bird is meager, and has rough feathers.

It is a sign that the blood doesn't circulate freely. Separate the bird, keep it in a dry place, give it 50 grs. iodine of potassium in 1/8 litre water, during several days.

Add also to this preparation 6 grs 25 sirop of gentian, 2 or 3 pieces of sugar as big as a horse-bean will be sufficient.

THE PIP

The pip consists in the drying of the inside of the beak and is caracterised by a white coloration of the corneous part of the tongue.

Sometimes little white vesicules looking like burns are to be seen.

This is due to a bad food and to a tack of hygiene.

Frist of all, alter the diet, and then if vesicles are observed, give astingent gargles of a liquid composed with alum, 10 p. c. iron-sulphate mixed in water.

CONSTIPATION

Constipation is as a rule due to a lack of exercice, provoked by a too small place, lack of vegetables, effect of diarrhoea, inflammation, intestinal gastritis, excess of astringent food, a want of water, etc.

The bird tries to expel the faeces ; it picks seed up, but stops every moment to renew its efforts.

It succeds only in expelling small quantities of hard faeces. The bird gets sad, downcast, its feathers become brittle.

Try heat treatment, vegetables, fresh water, and let the pigeon absorb each hour a tea-spoon luckwarm water wit sodium-sulphate 1 gram per bird.

If with this treatment no result is obtained, try to inject some luckwarm water with « Sunlight Soap ».

CATARRH OF THE CROP. — SOFT CROP. CORRUPTED CROP.

These diseases are caracterised by an inflammation of the mucous membrane of the crop. This ones swells seriously and gives the impression that the bird is wel fed, while in fact the crop contains but a colourless liquid or a decaying paste.

This affection is due, to the absorption of poison such as manure, bad seeds or owing to the fact, that a pigeon is unable to disgorge the food for its youngs.

In this case. the glands continue secreting the juice intended for the youngs.

This juice is not utilised ; accumulated it filters in, bet-ween the skin and the flesh, causing so a real corruption of the organs.

The ill bird is sad, secretes trough the beak a yellowish fluid of a nauseating smell. The feathers are bristling up, the eye is half closed, the crop hangs, the skin is violet coloured, thick, and feels soft.

The bird shakes the head and vomits.

As a preventive measure, always give first quality food in clean troughs.

To cure the illness, empty the crop by massaging it, the head of the bird hanging down.

The easiest method is to wash the crop with luckwarm water, to which some salt has been added, or with on. of the following solutions introduced in the oesophagus, by means of a rubber pear.

1 % Salicyl. acid.

1 % Chlorhydr. acid.

1/2 % Iron sulphate.

100 grs. boiled water.

After each injection, massage and empty the crop.

Repeat these injections twice or thrice in an interval of 2 hours, and give no food during 24 hours.

Feed then during a few days with bread-crumbs dipped in milk or puttered.

TYMPANITE OF THE CROP

This illness is characterized by an accumulation of gaz. The bird looks like a thick « Boulant » yard-pigeon.

Sometimes this illness spreads to all the cellular tissue, and the bird looks swollen up. In the most cases, it inte-rests the crop only.

The best way to cure this illness, is to prick the swollen part with a disinfected needle. Once the gaz evacuated, feed for a fortnight with seeds, sprinkled with sublimed sulphur.

PERITONITIS

Is often due to a perforation either of the rectum, or of the gizzard, or of any other intestine, due to the absorbtion of a foreign body.

The bird has diarrhoea, fever, and when the illness is chronic, it is always thirsty.

In certain birds, one can perceive several bodies, more or less hard, round shaped, circulating easily in the abdomen.

This illness is more often met with, in hens than in cocks, due to certain fragments of vitellus getting hard.

Empty the abdomen of its abnormal contents, and wash with a solution of potassium permanganate 1/2 per 1.000.

MICROBIAN ENTERITIS
GASTRO-INTESTINAL CATARRH

An inflammation of the intestinal muscous membrane is called enteritis.

Those caused by microbes, are more specially called microbian enteritis. These microbes live in the intestine and provoke an illness only when the membrane is inflamed.

It occurs often in tired birds, with have competed in many races without sufficient rest, and this specially at the moulting, in Spring or in Autumn.

The heat, the cold, the drafts to which the birds are exposed when transported in the baskets are factors injuring the gastro-intestinal secretions.

An imperfect food, damaged seeds, or seeds given on a floor soiled by feaces, can provoke enteritis.

When suffering from enteritis the bird is downcast, sad ; the crop swells and it is inactive, it has no appetite and it is bristly.

Its feaces are fluid ; then they take a greenish hue and provoke irritation at the cloaca where the feathers become sticky.

The bird gets meager, deperishes, is thirsty, worn out, and generally dies from intestinal hemorrhagy.

The bird will be separated from the rest of the team, and put in an even and soft temperature.

Its food will be easy to digest, such as boiles rice, breadcrumbs or linseed.

The water will contain 1 % iron sulphate.

Sprinkle the food with : 50 grs. pulverized ginger ; 4 grs. pulverized gentian ; 20 grs. pulverized quinquina ; 20 grs. fennel, anise, coriander ; 10 grs sulphate of iron.

A spoonful can be given in the morning, at midday and at night ; 2 grs. mint for 60 grs. water.

If the bird is constipated, give it castor-oil, or tea of senna (4 grs. senna macerated during 6 or 7 hours in 200 grs. water). This decoction will be given to the pigeons until they dung normally.

You can also give water to which 10 grs. rhubarb tincture have been added.

INVERSION OF THE CLOACA

Can be localised in the cloaca only, or in the last part of the bowels.

Its origin can be traced to too rich a food, to an intensive lying or to constipation.

Separate the bird, and clean the organ with a solution of potassium permanganate.

Rub it lightly and carefully. Introduce the finger, you have dipped at first in boric vaseline, just in the centre of the organ, and redress it combining with rotatory movements, pressings lightly forwards.

INFLAMMATION OF THE CLOACA

The inflammation of the cloaca is often due to enteritis or abnormal laying.

The membrane appears outside the organ and shows reddish, or even dark red. Sometimes it is even purple or black.

Remove all the little feathers surrouding the anus and put some boric vaseline or iodex on the diseased part.

When there is an ulceration, then give several times injections of lukerwarm water with 1 % iron sulphate. This will give very good results.

2°) RESPIRATORY ORGANS
CORYZA

Coryza, or a cold in the head, is one of the most frequent diseases amongst pigeons, which is often met in pigeon-houses.

As soon as the coryza manifests itse'fs, take the following precautions, to avoid it becoming chronic, because when it is chronic, it is difficu't to cure.

Separate the birds in warm and dry premises, and take care not only of these birds, but of the whole team which could be contaminated.

Clean the nostrils if they are obstructed. This is easily done by pressing the nose in the morning and at night.

Dip the nose till above the nostrils in a desinfectant. (Boric water 4 gr. boric acid in 100 gr. water).

To prepare the solution, add boric acid to boiling water and let then cool.

To wash the eyes, use a solution of boric acid, at 2 %, i. e. 2 grs. boric acid per 100 grs. water.

Regarding food, give the food that the pigeons prefer. As a supplement to the usual seed, give some bread added to milk, so as to make a rather thick pap. Twice a day give some rice with sugar.

As drink, give milk with sugar. From time to time, give water mixed with per liter 15 grs. iron sulphate.

Expose the birds in their resting place to fumigations. To fumigate use vegetal tar ; this will purify the lungs.

For the birds left in the pigeon-house, fumigations with vegetal tar, effected at night, give good results.

Of course the house must be thoroughly cleaned and also disinfected.

THE SNOT

The « snot » or contagious coryza is the most terrible disease. It is just a very dangerous cold and besides this very contagious.

The mortality encreases enormously if strict measures are not taken.

In the beginning the bird is sad, not lively, has no appetite, shakes its head, then gets diarrhoea, the feathers become bristly, the wings hanging low ; the bird sneezes and its nostrils are obstructed by a rather thick fluid, emitting a fetid smell and of a yellow greenish colour.

The tongue is thick, the eyes are tearful, bloodshed and the respiration is difficult till the pigeon finally dies.

Separate the ill birds as soon as you notice the disease and keep the other birds under close observation.

Their diet is rather special.

Spinkle the seeds with carbolic acid, 20 grs. pure carbolic acid in 1 litre water, and shake then the whole lot, to obtain a complete disinfection. In the first days, the pigeons will not take much of this food, but after a few days they will get used to it.

In the water given to the birds, add also 5 grs. pure carbolic acid.

Clean and disinfect thoroughly the pigeon-house.

The ill birds will be put in warm premises, and will be treated in a similar way.

Besides they will have a rather granulous paste, to which will be added a spoonful charcoal per 10 birds.

As drink, give them some milk with sugar instead of water. Twice or thrice a day, clean the nostri's, and remove the mucosities covering the inside of the beek and the eye, with a feather. Prepare a lukewarm boric solution, containing 20 grs. per litre, and dip the head of the ill bird in it, so as to wash the eyes the nostrils and the outside of the beak.

To wash the inside of the nostrils and of the beak, it is advisable to inject a little lemon-juice.

An other way to cure these pigeons, is to inject with a drop-counter twice or thrice a day the following solution: 30 grs. liquid paraffin ; 30 centigrs. menthol and 30 centigrs. camphor ; a pencil or a feather can be used for this purpose.

Fumigations of vegetal tar will daily be given to act up on the respiratory organs. Close hermetically the dove-cot. For this purpose, put the tar in an old pan, wherein you lay a hot piece of iron.

This cure is slow, but if one combats the illness from the first symptoms, good results can be expected with the hereabove given treatment.

There is also a contagious coryza. The birds scratch then their beak and nostrils, they sneeze frequently emitting a typical « tchic », but have never strong nasal mucus. They eyelids and ears instead of being dry and white are greyish, soft and slightly moist.

The inside of the mouth of a healthy bird must be pink, the split being well open, the palate must be of a nice pink colour, finely dentated in at its extremity.

If a bird is subject to chronical coryza, the mucuous membranes will be pale, livid, and covered with a greyish coat of mucus.

The tongue is white and furred, the split is closed.

The palate is congested, reddish or blueish, covered with thick mucus, the under part being swollen. Sometimes, its surface is covered with small white dots or yellow patches.

The symptoms are specially marked on wet or hot days, The disease is obstinate and cannot be cured by ordinary treatments.

After having used remedies which will conjure the illness for a short time, it will reappear.

So birds can be treated during years without obtaining satisfactory results.

This coryza is due to parasites (intestinal worms).

These ones are in the bowels of the birds, and the eggs are expelled with the feaces.

The birds pick them up again and develop them in great quantites forming so an unbroken circle.

The best way to help the trouble is to form a new team.

BRONCHITIS. — RATTLE

Bronchitis is an inflammation of the bronchia.

Bronchia are small pipes following the trachea and bringing the oxygen to the lungs, where it vivifies the blood.

This disease can occur under the influence of different causes :

1) Inhalation of irritating gaz, mixed or not mixed to the air.

2) Inhalation of cold air or damp touching the bronchia and provoking troubles.

The ill birds breath with difficulty, open their beak to inhale the air. Tthey keep quiet, strech their neck and the eyes are often closed.

This difficult breathing causes a rattle and sometimes nasal discharge of fetid smell.

Soon the bird loses its appetite, and if the affection reaches the large bronchia, the bird dies within a few days.

They get very meager.

Keep the bird in a warm place, well ventilated, far from any draft. You can also keep it in a well heated small room, and release tar vapours.

Let absorb sulphate of quinine, 50 centgr. a day and 1 gr. black sulphur of antimonium. Feed lightly. —

PNEUMONIA

Pneumonia occurs often after bronchitis, owing to over-driving, sudden change of temperature, etc.

The bird does not breath freely, conceals its head under the feathers ; the wings are hanging and the beak remains open.

Keep these pigeons in a warm room, and let them inhale vapours of eucalyptus leaves.

At night, every 2 or 3 days, introduce two or three drops of sulphate of copper or zinc in the nostrils, and inside the beak.

If the breathing remains difficult, pour on the back between the shoulders, turning aside the feathers, a fair quantity of iodine (25 to 30 drops).

THRUSH. — DIPHTERIA

Contagious disease.

Symptoms : small white dots, and then yellow patches. Ordinarily these patches adher only lightly to the mucuous membrane, and can be easily removed. Sometimes they are so numerous that the bird must open its beak, and make great efforts to breath.

As soon as these patches reach the outside of the beak, they prevent it closing, and the bird dies asphyxiated.

The patches are not always localized in the mouth ; they can reach the throat, the œsophage, and sometimes the abdominal organs.

The bird rejects soft and geenish feaces. It becomes sad, loses its appetite and the viscous liquid disgorged has a bad smell.

Disinfect the dove-cot, separate at once the affected bird. Twice or thrice a day, sprinkle the patches with soda bicarbonate, or scrap the yellow parts once every two days at night. Tent with a solution of copper sulphate or zinc sulphate (10 %).

Do not take this for an other disease, which appears like the thrush and which is easy to distinguih because its hard and thick patches are localized between the skin and the flesh, but not on the skin.

It is often met with in young birds descending from unhealthy parents.

Often it sits in the neck, on the crop, or on the abdomen or between the fork and the breast-bone. When localized on the abdomen, the bird is lost, and it is better to suppress it than try to cure it.

Located in the neck or on the crop, the operation is simple enough the bird does not even bleed.

Pull out the feathers on the ill place, and open to extract the contents.

Disinfect the wound with salt and if the opening is large, sew it taking the clasps away after a few days.

Do not allow these birds procreating.

THE WHITE DOTS ON THE PALATE

The white dots on the palate are not dangerous if not accompanied by yellow patches in the throat. It is only a sign that your pigeons are too ardent. In 1923, I have won many competitions while my birds suffered from it. The most important thing is, that the inside of the mouth be pink, and that the mouth split be free.

Give the birds marsh-mallow tea with couch-grass (tricticum repens) in equal quantities during about 8 days.

CONTAGIOUS DISEASES. — PHTHISIS

Phthisis is not frequent but very dangerous. The bird affected by it is sad, has pale eyes, dry and dull feathers ; the pectoral muscles are wearing away and the faeces are liquid.

Lack of hygiene, overdriving, weak blood are its principal causes.

Ki'l the i'l birds and disinfect the pigeon-hou se. If all
the team is ill, suppress it, incincrate them with all the
materials of the pigeon-house and disinfect this one.

3°) DISEASES OF THE CIRCULATORY ORGANS

Affections of the blood. — Anaemia.

An anaemic bird had dull feathers, pale eyes, the muscous membranes of the upper part of the mouth are bluish
and the extremity of the tongue is black, the bird has no
appetite and gets weak.

It finds its origin in a badly located house, damp, no
sun, a too long fasting, a defective food, overdriving, vermin, etc.

Observe hygien, and give the birds gravel, crista!lised
sugar, salt, phosphates, oyster shells and abundant greens.

Add to the water 10 drops of the following mixture :

10 grs. ammoniacal lemon acid of iron. (iron citrate) ;

3 grs. Fowler liquor ;

85 gr. Fluid extract of sarsaparilla, (smilax medica).

RACHITIS

Special to young birds and can become chronicle. This
il!ness slackens the growth and the normal development of
the bones. The cartilage of the young does not ossify properly.

The ill bird has rough feathers, no appetite, liquid faeces. it is weak, walks and flies away with great difficulty.

Badly exposed or ventilated house, lack of hygiene, privation of chalk, gravel, sugar, salads, etc., can provoke this illness.

Observe a strict hygiene, add to the usual drink 5 or 6 pieces of sugar, or a spoonful of pure honey per litre of water.

WINGS' DISEASE

(Articular Rheumatism, Arthritis, Lameness)

The wing works badly and the disease interests either the nerves, muscles, or the joints. As a rule the joints are ill.

The bird suffers much and this in an encreasing way, this compelling it to carry the wing in the less painful position.

Ordinarily the wing hangs, the feathers are rough, the damaged parts become purple.

This coloration is accompanied by high temperature all around the ill part, and when you feel the bird, it shows that it suffers violently.

The origins of the disease are numerous, the principal being heredity. — Other causes are fatigue, young age, bad weather, rainy weather, confinement, repeated efforts, etc.

The following cure is advisable :

' Massage on and under the joint during 6 or 7 days, with a mixture of :

3 grs. henbane extract. ;

10 grs. Roputerum salve ;

25 grs. Mercurial salve.

If the case is a bad one, give during 2 or 3 days a spoonful of disti.ated water, 100 grs,

Fowler liquor, 1 gr. ;

Iodium of potassium, 2 grs. ;

Salycilate of soda, 2 grs. ;

Shake the bottle before using.

Keep the bird at good temperature, out of any draft, place little rods to help him climbing up and down. The bird must be kept moving.

Twice a day, immerce the bird in lukewarm water and keep it so, for several minutes. During the bath, open and close the wing, seeing if it works well in every way.

4°) DIFFERENT DISEASES
SMALL POX

As variola for the people, this disease can attack either the surface of the skin, or the outside of the mouth.

It finds often its origin in a bad hygiene, dirty water, too small pigeon-houses ; sudden variations of temperature can also bring it about.

The illness begins by sadness, fever, a lack of appetite, ruffled feathers, dry and bristly quills.

Then after a few days, you will observe around the eyes, ears and beak, as well as in the inside of the thighs small pustules, encreasing in size, and sometimes gathering in a mass.

After this phase the pustules get open emitting a yellow fluid drying a short time afterwards. From this moment on the bird gets merrier and wins strength.

When these pustules are in the mouth, the rear of the mouth, or in any of the digestive organs, they attack the respiratory organs, the digestive organs, the nostrils, the larynx, the bronchia, and sometimes provoke pneumonia or a serious bronchitis.

1) On the surface of the skin, the following mixture which does not injure the eyes, will be applied.

2grs. Arsenious acid. ;

2 grs. Powder of cantharides ;

5 grs. rich turpentine ;

8 grs. wax ;

15 grs. oil.

2 grs Arsenious acid ;

You can also rub a little milk of piss-a-beds or dandelions over the eyes. Take the whole plant, with roots and leaves, press then the radix, and the milk will come out.

In the mouth the illness is contagious and the bird which has this illness will be separated from the team and the pigeon-house must be cleaned and disinfected. The bird will fast during 2 days and will drink 100 grs boiled water added to a piece of aloes, as big as a Scotch bean.

After having fasted, the bird will have some buttered bread with some garlick and fresh salad. As drink, give it water with some sugar or honey.

Anti-diphterite vaccination is advisable.

This should be done when the birds are 6 or 7 weeks old, considering the fact that the pigeons get this illness when they are in the hampers, and that the young birds are more subject to it than the old ones.

Since the « Roux-Behring » serum, vaccination is easily done by a veterinary.

INFECTION OF THE NAVEL

Illness of young birds.

It occurs a short time after hatching. The birds which are subject to it seem strong and fit, but they die a few days after their birth.

The navel gets contaminated by contact with the nest. Before pairing the birds, clean thoroughly the cells with milk of lime.

The nests intended for breeding will be provided with a layer sand or saw-dust and live lime in powder.

Cleanliness is very important.

When the new born birds are well dryed, apply in the wound with a pincel some iodine glycerine. This can cure the disease.

CHAPS

The legs can have chaps and suppurate.

It occurs in birds resting on a wet ground or when cut or wounded.

In this case, observe strictly the hygiene and keep the house as dry as possible.

Cover the chaps with the following mixture.

Silver nitrate, 2 grs ;

Boiled water, 50 grs ;

One can also use vaseline, 25 grs. ;

Perou balsam.

INFLAMMATION OF THE CONJUNCTIVA
HERATITE. — CATARACT

The inflammation of the conjunctiva is an affection of the membrane of the eye and of the internal part of the eyelids.

Its origins are numerous, and the most frequent beside traumatism, are diphteria and contagious coryza.

This illness without any complication, is often caused by a stroke or by a foreign body in a badly kept dove-cot where nails or other things are protruding.

1) In case of diphteria, this inflammation of the conjunctiva is uncurable. The eye becomes watery and pale ; the eyelids are moist and dull. One can notice a small node, as big as a rape-seed grain.

2) In case of coryza, the eye is dull, moist, the eyelids are swollen, reddish, nearly always closed. A clean greyish, sometimes yellowish fluid is emitted. This one dries and sticks the eyelids together.

3) Cataract is often caused by parasites of diphteria. First of all, the reason of the illness must be found.

If it is caused by a foreign body, try to extract it and tfen apply a soothing mixture.

If caused by a stroke, showing a wound, apply an astringent and soothing ointment, which can bought in any pharmacy.

If caused by diphteria or coryza, do not treat the inflammation of the conjunctiva, but the real illness.

As soon as the cause is suppressed, the effect will disappear. Each time a bird suffers from the eyes, it is advisable to wash it several times a day and introduce 2 drops of the following mixture.

50 grs. Rose-water ;

2 centigr. Zinc sulphate.

If spots appear on the cornea, use the following eye-water :

Boiled water at 212° Fahrenheit, 50 grs. ;

Silver nitrate, 25 grs.

Let fall 2 or 3 drops in the eye in the morning and at night. If the mucous membrane is attacked, go then as soon as you stand up in the morning, before the breakfast to your pigeon-house and put some saliva on it.

From time to time, let fall a few drops boric water with some ground sugar.

STIFF NECK

In certain pigeon-houses, pigeons with stiff necks are very often met with. It is characterized by troubles in the balance of the muscles of the neck, compelling the ill bird to carry its head on a side.

This happens often when the breeder gives a food which is too rich in substances containing azote, or when the bird does not drink enough or when it has been overtired by a long flight during a very hot day.

This disease is very difficult to cure.

At the beginning, a bloodletting by cutting the nails can have an immediate result. The ill bird will drink fresh water with bicarbonate of soda, 10 grs. per litre.

It will be given minced vegetables, lettuce and tender herbs.

If the bird is not healed up after three or four days, pull out the feathers on the opposite side towards which it turns the beak, and rub with tincture of iodine (tinctura iodei).

POISONING IN THE FIELDS

This poisoning is due to the artificial manure spread at certain periods on the fields, viz. in sowing-time.

As soon as a bird, having absorbed this manure comes back to the pigeon-house, it rolls itself up in a ball ; it is sad, has diarrhoea and is very thristy, every organ being inflamed.

After this, wash the crop with a rubberpear filled with boiled water in which you have put some salt.

If te condition improves, purge the bird with 50 grs. boiled water with an aloes grain as big as a horse-bean.

The following day, give it as drink some pure milk, and as food 10 to 15 small balls of fresh bread with some garlick-heads.

DIFFICULT LAYING

First let the bird vomit, by keeping its head downwards and press the crop, so as to have its contents evacuated.

Some birds lay eggs with great difficulty. This occurs often by felales which have been kept in captivity for a long time, or by yearlings of which the pelvis is not well developed.

An inflamed oviduct, (passage for the egg from the ovary to the external outlet), or constant constipation are its usual causes.

When the bird has an egg, choking the oviduct, lays on its back, has no appetite and rolls itself in a ball, it must be well observed. From time to time, and not at the usual hours, it tries to expulse the egg. Sometimes it emits a thread of blood by opening the cloaca which is hot, purple, red and tumefied.

Keep the bird warm, in a basket provided with some straw.

Anoint the rear part of the oviducte with some olive oil and dilate the cloaca by means of fine oil. (olive oil or an other some).

Give them also a lukewarm bath every 2 hours.

EGGS WITHOUT SHELL

The egg has no shell and is simply covered with a membrane.

This is due to a lack of liberty. When the bird flies out, it finds lime-stone, etc., what it wants.

Give it pieces of bricks, gravel, crystallized sugar mashed vegetables.

5°) PRINCIPAL PARASITES OF THE PIGEON ITCH DEPRIVING THE BIRD OF ITS FEATHERS

Contagious disease.

Very serious illness. The pigeons shed their feathers in any season.

The parasit attacks the root of the feathers and cuts the lower bristles of the primary quills.

Clean the cot as soon as possible and give twice a week a sulphureous bath to the whole team.

ITCH

Parasitic and contagious disease due to the itch acarus.

Appears on the skin, not cured in due time, and if can take a considerable extent.

Any il' bird shows a short of whitish dust encreasing and turning to greenish yellow.

Wtihin a few days, a great number of vesicles appear causing severe itching on the whole body.

These scabs stick very hard to the body, specially to the legs, so that they distort completely their shape ; if removed, they can even provoke hemorrhagy.

Its cause is a microbe called « sarcopte ». The disease can be avoived preventively by giving once a week a bath, (water with a handful of salt).

Anoint then the ill parts with the following ointment.

5 grs. Salycilic acid ;

5 grs. Precipitated sulphur ;

25 grs. Vaseline.

Apply each day until your pigecns are healed.

Disinfected the place and the materials that the ill birds have used.

WORMS

They are often met with in great number in the digestive organs.

The bird is meager, downcast ; it as always hungry, and when you go to the pigeon-house in the morning before the bird has moved, you will notice hundreds of small worms in the feaces.

Separate the ill bird, and remove the feaces as often as possible.

Give fresh water to drink (100 grs. with a piece of aloes as big as a chestnut).

In the morning before the first meal of the bird some balls fresh bread with garlick in equal quantities.

As food, linseed and flat millet will be given in equal quantities too.

After the disease has been cured, clean thoroughly the house and give the bird two pieces of sugar as big as a maize corn mornings and evenings.

LICE, FLEAS, ACARES

These are well known by the breeders, and should never be met with in any house.

These insects prevent your birds from resting properly. They multiply and disseminate themselves extraordinarily rapidly.

The causes which bring about insects are numerous. The lack of lukewarm water bathes and the transports for the competitions are the most frequent ones.

Of course the house must be thoroughly disinfected as well as all the material.

Fumigations by formol or by burning sulphur are very efficacious.

Plunge the birds up to the eyes in a lukewarm water bath during a minute and spread under the feathers a mixture composed as hereafter specified.

30 grs. Sulphite of potassium mixed in equal quantity to Spanish camomille flowers per litre water.

It is advisable to add an insect killing preparation.

A wooden vessel will be used, sulphite reacting with metals.

Therefore it is better to use a decoction of « Quassia Amara », suriname wood, in order to kill, the parasites.

This wood is used in chips, which are boiled during 10 minutes, 10 grs. per litre water.

Saphisaigre or lice herb can be used, spread in the pigeon-house to kill the lice, as well as all the parasites.

This is the secret to always have pigeons free from vermin, this being quite nefarious.

Special Parasites, to be noticed en the stems of the feathers as the microscope has shown me.

Alcohol is quite useful to keep away those parasites. Take a pinced and cover alcohol each infected stem. Repeat it several times.

RED PARASITES

Are dangerous for they suck day and night the blood of the pigeons. They are very voracious.

They are due to old buildings, boards and frames with large openings, badly exposed premises.

All their nesting places must be destroyed. Remove the team during a fortnight, whitewash the house twice a week and let the draft have free play.

PIGEON LOUSE OR TIQUET

Sometimes mistaken for red parasites, it has an oval flat body ; sometimes narrower on the front part. It has a chestnut brown colour and is 5 to 8 m/m long. and 3 to 4 m/m broad.

It lives in the vertebres and sucks the blood of the pigeons.

The abdomen swells in such a way, as to become 10 times as big as in the normal state. When fasting, it looks like a seed and can live 3 years in a sealed vase.

In the pigeon-houses, it attacks specially the young birds and can even provoke death by exhaustion.

They fix themselves at the breast and at the neck when the birds are at rest.

Special traps are in the trade to catch them. They are fixed under the nests of the pigeons, and provided with a special gum, attracting these parasites and in which they stick.

This is a good way of dealing with these parasites. One can also chase them and the straphisaigre or louse-herb turns them away.

The splits in the walls must be avoided for the lice ; they lay their eggs there and these are very numerous.

The tiquet broods during 19 to 20 days. One of the most efficacious way of getting rid of them, is to prevent them brooding.

ACCIDENTS
OBSTRUCTED NASAL CAVITIES

Caused by a foreign body or by fibrous deposits due to a stroke on the bill, etc.

Extract the foreign body or fibrous deposits and wash the cavity with a solution af permanganate of potassium.

OBSTRUCTED TRACHEA

The trachea can be obstructed by nails, pins, iron wire or seeds.

The bird can hardly breatch, shakes its beak and head.

Open its beak, and examine every spot. Feel the crop ; this one can be opened to extract the foreign body. Before allowing the bird to feed, wash thoroughly the crop with water containing some salt.

BLACK SPOT

It can happen by pulling the feathers out, by picking at the root of the feathers, by wounded feathers, or by a bad diet or a too great fatigue.

To cure this illness, give your pigeon twice a week a bath with lukewarm water and a tonic food. Thereafter it must rest till the damaged feather is entirely reneved.

OBSTRUCTION OR LACERATION
OF THE CROP

I mean here a laceration provoked by a telephonic wire, or due to a hawk, or an obstruction caused by a large foreign body, preventing the food to enter the stomach.

If the crop is obstructed, make an incision in it, after having cut the feathers all around the wound.

Once the foreign body extracted, wash the wound and sew with silver thread or aseptised thread.

If you observe a laceration, remove all the feathers around the wound, wash thorougly and sew the wound.

Feed lightly during 2 days, and give your pigeon 2 pieces of sugar as big as a maïze-seed, morning and evening.

RUPTURE

Ruptures are rather frequent amongst our pigeons ; specially in the houses where there is no external landing board, or which are surrounded by electric cables or telephonic cables, which are not very visible for the birds when the sun is shining or when there is mist.

When the bird comes back with a rupture, the accident is not very serious as a rule, for a seriously wounded bird falls and is caught by cats or an other enemy.

Reset the bones carefully and fix them with plaster.

If there is a wound, disinfect before dressing.

The dressing plaster may be removed after 8 days.

During the treatment, sugar strengthening the bones will be added to the food and to the drink.

WOUNDS

Consequently to chocs and lacerations, it happens that the bird is not able to take the air again.

In this case, separate the bird, clean the wound, and apply oxygenated water.

PRECAUTIONS TO BE TAKEN IN CASE OF DISEASE

Very much attention will be paid to the hygien and to the disinfection.

Give to each diseased bird 25 grs. boiled water with a piece of aloes as big as a horse-bean ; this to clean the body from any poison in the system.

Apply one of the methods described for the various diseases, as the case may be.

If the disease happens in several birds, take the necessary precautions and ascertain if this disease is not due to a lack of air, of vitamines or of sunlight.

Therefore I once more draw your attention to the fact, that your pigeon-houses must be rightly set, and must get as much sunlight as possible and must always contain salt. Never forget to distribute chopped fresh green once a week during the season of the races, the day following the race.

CURE AT THE END OF THE SEASON AND IN SPRING

Give the bird a purge.

A few days before giving the birds this purge, it is advisable to feed freely with linseed, in order to prepare the bowels. The day before, you will give a meal only of these seeds in the morning suppressing the drink. Of course, the pigeon-house is to be shut, in order to avoid the pigeon to go and drink somewhere else.

The following day at about 10 o'clock, give the purgative drink, when the birds have had no food.

The drink will consist in 50 grs. English salt or sulphate of sodium per litre water.

Leave the drink in the house for a few hours, and when 1/4 only is left, add some water to maintain purgative properties.

At about 4 o'clock, give half a ration of ordinary food, mixed with linseed. Supply linseed during a few days still diminishing progressively from day to day reverting slowly to the usual food.

Secrets and Informations

QUALITIES OF THE REAL BREEDER

To be a good breeder and to be successful in this fine sport, certain qualities are required, viz. gentleness, perseverance, observation and reflection.

GENTLENESS

Everybody will admit, that neither from a pigeon nor from a man, nothing can be gained by violence or rudeness.

A pigeon by its own nature is timid, and it must get used te its master to become its friend.

An abnormal noise, an unusual attitude or an abrupt movement scares the bird.

Never enter the house without warning the birds either by your voice or by whistling, and take advantage of the distribution of food in the individual troughs to familiarize them by talking to them; when going along the nests caress the birds, even the most timid ones.

When the distribution of the food takes place, go in the middle of the house and compell the birds to be familiar with you, in order to make them docile.

Be kind, caress the birds, let them pick up the grains out of your hand and have nothing to scare them.

When you must catch a pigeon, bring it back to its nest where it is less shy and where you will have no risk of wounding or damaging its wings.

You must try to make yourself understood by the bird. Never catch it without talking to it. Make it familiar with your voice, your attitude, etc. ; a timid bird is a nuisance in a pigeon-house.

PERSEVERANCE

Be firm and perseverant. Never let you lead astray from your goal by other persons. Habits of a good breeder are difficult to acquire, but after having studied this book, you must be able to direct your team, and reach your goal without needing advice from anybody.

Introduce no alteration neither in the diet, nor in the training.

Be firm, for if you have adopted a method and if you have confidence in it, you must follow it, even if your birds are not successful.

It is not by altering the method, that you will improve the results.

Be perseverent if your birds are healthy, and get good rood ; at last, they will be successful.

OBSERVATION AND REFLECTION

All the science of a breeder is comprised in these two words.

To be successful, you must be a keen amateur, but also a keen observer. Only by numerous observations, you will be able to understand the character of each bird and to realise if the presence of the mate has a good or a bad influence on the subjects.

The instructions given in this book are useful, but they do not exclude your personal observation. During the travelling season, when you have some time to spare, go to your pigeon-house and observe the birds one by one.

In the chapter reserved to the various systems, widowhoofd, half-widowhoofd, etc., I told you that when you wanted to observe the birds, you had to go noiselessly to the pigeon-house and watch them through the pane in the door or in any other spot, affording a full view of the house.

When you enlist the pigeons or put them into the hampers a few hours before the start, let all the birds take the air, and specially those intented for the race.

A bird which was in a perfect condition yesterday can be ailing to-day.

Wings trouble can occur suddenly, and a single night is enough to place the bird in a poor condition.

For this reason you must observe the birds as much as possible.

By observation and knowledge, you will be able to get excellent results.

This point is very important, and I shall be satisfied, if I have sufficiently impressed on you all its importance.

STRICTNESS AS TO THE SELECTION

One of the principal causes of lack of success, is the bad choice of the birds.

No success will be possible as long as you have not se'ected a founder of a family or better still a family, perfected by your care and endowed with the qua'ities preferred by you, which will be bestowed on the offsprings.

Every pigeon is not apt to procreate, to travel and to compete.

This is the reason why certain qualities are needed to stock the house with birds all suitable for flying and procreating.

No success will be obtained if you introduce each year new individuals in your team, and if you do not limit the number of your birds to a quantity permitting a good inspection of each bird.

Keep only good birds, avoid even moderately good ones.

As soon as the travel season is opened, birds getting old, and those ones which have not given good results neither as carrier-pigeon nor as procreator will be put aside. Do not wait for the end of the moulting-time, it will be a loss of time as well as a loss of money.

Amongst the young birds of the year, do not take into consideration the fact that the youngs are from a famous stock, but look out for birds having inherited the qualities expected from this particular stock. They must moult regularly without trouble, and develop nicely. The organs must be fortified by the development of the breast.

If these qualities are lacking, put these birds aside.

NEGLECTED HYGIENE

Lack of success is often due to a bad hygiene. Many houses are not sufficiently exposed to the sun and also badly ventilated.

Others are subject to drafts.

Others still are badly set and instead of getting pure and fresh air, sunrays and all its ultra-violet rays, they are moist and the colony suffers from numerous diseases.

Often the cleaning and disinfection are neglected with the result that dirt accumulates and causes the development of parasites and microbes.

Water plays also an important part in the hygiene. This is the best medium of development for microbes.

The majority of contagious diseases are due to water, given in dirty fountains or troughs.

FOOD

Food must be carefully selected and must fulfil the requirements already specified.

Lack of success can often be traced back to a bad or badly distributed food.

The breeder often spoils his birds by too rich a food.

Certain breeders do not give the weekly bath to all the birds, and do not provide the house with the condiments needed by his pigeons.

Regular training is also necessary for a subject to be fit.

In any kind of sport, a regular training is the basis of success.

A subject which during a certain time covers short distances or which is kept idle, will be exhausted when it must travel.

If your birds fly each morning and night during one hour, they will not feel tired when having to cover an equal distance when travelling. The rest of the distances is also easily covered when the subject is fit.

It often happens that breeders complain of having no success anymore, with birds, which the previous year were quite successful.

Having no method, knowing nothing about the sport, they consider themselves unlucky when they should attri-, bute this lack of results to their ignorance. When they possess one or several good subjects, obtaining good results in competition, they are unable to moderate themselves and start the birds on every opportunity without taking into consideration, in which condition the birds are, and these ones get worn out in no time, having no rest at all.

The real breeder starts a bird only when practically sure of obtaining one of the first places i. e. when it is in perfect condition.

BAD « HOME COMING » FROM THE RACES

Of course no alteration will be brought to the entrance of the pigeon-house during the travel-season.

The origin of bad home-coming is often the lack of tact of the breeder. These bad returns from the races are often met with by timid subjects.

When a bird comes back tot the house, do not go in at once to take off the competition ring, be patient and let it enter its nest. Never take a bird which comes back from the race, out of the entrance of the pigeon-house if it is not used to that.

After having familiarised a bad returner, if it goes on landing on the roofs when coming back, place a crate where the pigeon lands usually and put it alone in its crate, for 2 or 3 hours until the daily flight. Take it out then, and let it fly with the other birds. Repeat this punishment until the bird feels an aversion for it.

When it will have been locked up alone during several days, carry it to about 100 M. from the house and release it there.

If when returning it does not drop at once into the en- trance, put it back in the crate and punish it so until has lost its habit.

The same rule is to be applied to a bird which keeps flying above the house instead of entering the house at once ; but instead of being put on the roof, this bird will be kept in freedom alone and flying until it comes and hangs to the entrance trying in every way to get in. This will be a proof that the punishment has been successful.

BAD MATING

Mating is not always effected in ideal conditions, many breeders do not take enough care of it.

The method is explained chapter 3. Avoid as much as much as possib'e changing the cock or the hen during the period of the competitions.

Any pigeon is attracted to its mate and does not leave it unless compelled, and it happens that, trying to mate a pigeon with an other birds the results are spoiled : the pigeon does not show the same impetousness and often it turns naughty.

Be careful, and mate birds of suitable dispositions, able to improve the courage of their mate for coming back from the races.

IRREGULAR MOULTING

I told you that the feathers of the pigeon are like a book, telling the state of health and the dispositions of the subject.

Pigeons which have not moulted regularly show the following defects :

1) Transversal stripes on the stems of the primary and secondary quills or rear-feathers with openings or splits.

2) Fine down exceeding in length the ordinary feathers on the crop and on the neck.

3) Thick down under the wings or around the rump.

4) Rough, dry, bristly feathers.

5) Parasites on the stems of the primary feathers.

When inspecting your team, reject as unfit for travelling all those, presenting one of these defects. They will never get results until they have undergone a certain rule of diet and until they have had a good moulting.

An easy way to see if the pigeons have had a good moulting, and to guess those which will get success in the following racing season is, to enter the house when it has frozen during three or four days and to inspect your birds one by one.

Those which will have quills undulated to the rachis will be classed as unfit.

CHANGES IN THE PIGEON-HOUSE AND LACK OF CARES AT THE « HOME-COMING »

The pigeon must never be annoyed, specially during the racing-season.

Close the pigeon-hole of the bird when it is away, avoiding thus that an other one takes it. Both birds would fight at the return and this would be detrimental.

No change must be brought neither to the entrance, nor to the outside of the house ; wait to carry them out till the travelling season is over.

When coming home from a race, the pigeon can bring back an epidemic contracted in the baskets, and only one contaminated bird can spoil the health of a whole team.

POPULATION OF THE HOUSE

Successful houses do not always contain a great quantity of pigeons, as a rule they contain only a few birds, but these are picked out and carefully looked after.

A crowded house is always poor in success.

CHOOSING THE BIRDS

Pick out your young birds and have only subjects fit for flying.

This choice will be made severely, for on it depends the future of the colony.

First you must judge the bird by handling it. The subject must be dry, hard, nervous and sanguine. Never choose a soft, lymphatic bird.

Observe the development of the chest. This one must be wide, containing well developped lungs and air pouches.

Examine also the pectoral muscles make sure of their power and observe the way in which they are spread on the bones.

Choose subjects having a strong breast-bone, pointed, slightly curved forwards.

Then open the wing, measure its span, control its flexibility and the stregth of its attachment.

Make sure that the muscles of the forearms are strong enough.

Observe also the primary quills to see if they are fine and flexible.

Open the wing, turn the subject on its back, and look if the pipes of the rachis are white and well assembled.

The feathers must be rich and velvety.

The head must be strong with a small flatness on the upper part ; the forehead will be high and broad with a light swelling at its top.

Choose the blasckest pigeons.

Your pigeon have a narrow but not too long tail.

The neck must be of a very strong structure.

The fork must be strong and well fixed at its extremity.

The legs have a middling height.

ADDUCTION OF YOUNG PIGEONS

All the amateurs must possess a crate so as to be able to place it on the roof or in the neighbourhood of the house when there are young birds or when a new bird is to be adducted.

If the youngs are pigeons of that year, as soon as they eat alone, place them in this crate during a few hours by sunny weather without letting them out.

When they are in the crate, give them seed and fresh water, and avoid that the old pigeons trouble them.

They will then begin to examine the sky and the neighbourhood.

After a few days spent in this crate, let them out when the wind is not strong and after the evening-flight of the old birds.

Ordinarily old pigeons lead the young ones too far away, so that they get lost.

The best way to proceed, for those who have a fairly large house is to wait for the moulting of the first quill on each wing before letting the young pigeons out for the first time.

Give them then the required condiments and chopped greens.

ADDUCTION OF STRANGE PIGEONS

Many breeders find it hard to adopt and get accustomed a strange pigeon.

One must dispose of a rather large space near the entrance of the house and place the bird there with its mate, choosing the most convenient weather te release it.

If by its former owner it has got success or if it has procreated good subjects, mate it with a bird exactly similar to its former mate.

Be it a cock or a hen, always observe this rule. If the bird is a male, wait until it wants its hen ; then let it go out with her alone for a few flights.

If you have to deal with a hen, wait to set her free that the eggs are craked or that the youngs are newly hatched and keep the male away the whole day of the adduction.

This method for the cock as well as for the hen, must be followed until the bird has got used to its new house and to its new companions.

You will also choose a favourable weather, for when the weather is cold and misty, the birds do not find well their way, and sometimes they are obliged to stay out the whole night.

CHANGING FROM PIGEON-HOUSE

When you have to change the pigeon-house for the whole team, the task is not difficult, and the distance does not matter.

To have full success, begin a fortnight or 3 weeks before changing.

Separate both sexes and dismount all the fixtures, so that the birds are compelled to remain on the floor.

Fix the new house as the old one was.

Once the new house is ready, carry there all the hens ; fix the cage on the roof or on the landing board and have the access easy in order to allow them to examine the neighbourhood.

Take care that the hens take possession of the same holes as those they had previously in the other pigeon-house.

When the cocks are moved, they will at once go to their hen and settle down.

After 2 or 3 days of confinement, as soon as they have got used to the new house, they will soon have observed the neighbourhood.

As to their first flight, wait till they build their nests ; confine all the hens in the holes ; leave the cocks free and if you have closed the old entrance of the first house, if even they go back to it, more than the half of the team will come back to the new house.

As to the adduction of the hens, wait till they brood since several days, or that the youngs are freshly hatched.

If you follow this method, you will not have to fetch your pigeons back twice.

EASY AND QUICK ADDUCTION

To adduct some subjects, give them :

2 litres rain-water ;

500 gr. washed hemp of Chili ;

5 gr. Lavender flower ;

5 gr. Essence of anise.

Boil this mixture during 5 or 6 minutes.

Strain and let get cool.

During 2 or 3 days give this liquid as drink and the seeds as food.

During this period, after the evening meal, give a piece of old cheese or even better a piece of horse-liver fried in butter (the piece must be as big as a horse bean).

Before setting free, confine in the holes a good part of the team and wait until the evening flight is done.

Certain amateurs use the most extraordinary mediums, which I have never tried. I can assure that the smell of certain plants such as : caraway, cumin, hemp, anise suit the taste of the pigeons, which are drawn to them by their emanations.

These smells can be advantageously used, and certain amateurs sprinkle the surroundings of the house for a quick return during the races.

HOW TO CONSERVE PIGEON EGGS

To keep the eggs during 10 or 12 days, before giving them to a breeding couple, considering that the first egg is never brooded before the second one is laid, take them away from the nest as soon as the second is laid, and place them in a vessel containing a layer of 5 cm. bran separating the eggs from the air.

Turn them round once a day, without exposing them, in order to keep them their qualities.

Before giving them to be brooded, place them a few minutes in a cup of fresh water ; dry them up and put them under the breeding couple.

They will hatch within 17 or 18 days.

NO EGGS LAID DURING THE RESTING PERIOD

The amateur which has only one room for his pigeons, and which does not want his subjects to lay eggs during the resting period, will proceed as follows :

In 10 kgs. seed, he will put a piece of camphor as big as a chestnut. Shake the lot twice a day, keeping the camphor in the centre. After a week, give this food.

By this method, the pigeon retain their strength, moult well, and through well fed, will not lay an eeg as long as this system is followed.

FECUNDATED EGGS

After one or two day's incubation, you will be able to discern if an egg is fecundated.

Take the egg between the thumb and the forefinger on its length, hold it up to the light, and you will observe :

After 2 day's incubation a line and a spot.

Following day, two small lines will appear on each side of the first line.

The 4th. day, though everything is embedded in a liquid, you will easily distinguish the head and the heart.

WILL THE GERM GIVE A COCK OR A HEN ?

As a rule, the form of the egg reveals if the embryo is male or female. A round one gives a female bird, and an oblong one gives a male bird.

By the male, the germinal spot in an egg is nearer to the extremity than to the center. By the female it is the contrary.

CARES TO BE GIVEN TO OLD PRODUCERS

I am often asked which cares should be given to old producers for obtaining good youngs from them.

Give them during a period of rest before mating, horse-beans germinated in wet earth.

During this period, give them 2 or 3 pieces of sugar, a week, as big as a horse-bean. This will be given in addition of the general cares to be taken of the reproducers.

ABNORMAL FEATHERS
BENT QUILLS

When the stem of one or several primary quills or rear feathers of a bird are misformed, or bent without being broken, dip them during about I minute in lukewarm water; then give them their normal form by pressing with the fingers. Once this operation is performed, let the bird dry before releasing it.

BROKEN QUILLS OR REAR FEATHERS

Replace them as follows.

OPERATION : Wrap the pigeon up in linen cloth in order not to damage it and to keep it quite.

In order to case the pipe, cut in bevel the damaged feather at the place where the barbles begin.

Take then a feather from an other subject, smaller than the feather in question, which you will have put aside, numbering it during the moulting time. You can collect those feathers.

If e. g. the bird has broken the 5th. feather, beginning by the outside, take a 4th. one of the same wing, cut the point, and cut a bevel opposite, to the one of the feather you have just cut.

Apply some glue on the outside of the 4 th feather and on the outside of the 5 th. one.

Introduce the feather into the shaft of the other feather.

Open then the wing and close it several times in order to make sure that the feather is correct in size and fits well.

Keep the subject wrapped up until the feather is dry. The new feather will be quite strong.

HOW TO RECOGNIZE THE SEXES

A cock can be distinguished from a hen quite easily when you deal with an adult.

As a rule, the male has a bigger head and throat, is generally bigger, has well developped caruncles, shorter fork bone, extending more directly without presenting a sort of hole as soon as the finger leaves the breast-bone.

The quills are narrower, more pointed at the extremities. For certain shades, the sexes are shown by different signs.

E. G. — Red birds and pale ones with one or several black spot on the quills or on their tail feathers are cocks. Hens have no spots or only rusty coloured ones.

For carrier-pigeons, a difference permits to discern practically surely the sex as they are ar least 120 days old. The caruncles have a white line separating both swellings in the hens, which does not exist by the cocks.

For young birds in the nest, the cock has the biggest bill.

To be certain of the sex of a bird between the moment of leaving the nest and the moment they become adult, turn the bird on its back in the left hand, open with two fingers of the right hand the lids of the cloaca and observe this organ.

By the cock, you will notice 3 conical convergent protuberances, a dorsal one in the medium line.

By the hen, 5 convergent furrows (groves), in star, one being median dorsal and separated by as many protuberances.

EXHIBITION PIGEONS

Every bird does not possess all the qualities required for an exhibition.

Choose a well balanced bird, having rich and abundant feathers, corresponding to the standard fixed in chapter one.

Cock or hen, the head must be round, the caruncles smooth, the eye dark and lively, well set, fine membrane surrounding totally the eye, strong neck of medium length, well developped chest.

Wide and broad wings are required as thick mantle, covering entirely the rump, and narrow tail of medium length.

During the exhibitions, which take place as a rule during the resting time, the subjects will receive freely a heathing food, similar as mixture to the one intended for birds entered for speed contests.

In order to give them the port and carriage required, place the bird in a crate, out of the house and in a rather crowded place, and repeat this twice a week during half a day.

The day of the exhibition, keep the bill and legs quite clean.

SPECIAL PIGEONS

Every amateur, must have one or several trained pigeons in his team, for the best subject returning regulary as a rule, can some day let you lose a lot of money, by flying in the neighbourhood of the house, or landing on its roof.

This can happen if something abnormal scares it and it will then sometimes return only after the competition is closed.

A pigeon intended to attract its mates in the house must not be entered for competitions. During the travel season, it must be kept just a bit hungry and jealous.

To get it used to return quickly, at feeding time, before distributing the food, put it out of the house and place an other bird in its hole.

Give this last bird its ration in a small trough and if you play the natural game, add in the same trough the ration of the hen. Wait till the colony has absorbed the main part of the ration before opening.

If this method experienced with several birds does not give a good result, try the following system giving also very good results.

Take an old bird, by preference a hen. Cut the four long quils in one wing on about 5 cms., and take the bird to about 20 M. from the landing board.

Release the bird, and if it keeps flying about, cut some more the same feathers until it has some difficulties in getting back to its house.

This method is excellent, but to have a compelling bird making a quick return, it must be used only at the last minute at about 150 M. from the house.

INFLUENCE OF WIND AND WEATHER

When you examine a team, you can discern, taking into consideration the physical conformation, from which the moral depends, how each subject will be successful according to the fact, that it has the wind either ahead, or astern, or aside, and that the weather is clear or covered.

Every pigeon is not affected in the same manneer by the wind or temperature.

A flat and short headed individual with yellow correlation circle and primary quills, pointed, slightly turned will be successful by clear weather and by South or South-East.

An individual flat and short headed also, with a correlation circle, yellow or white, or even without any correlation circle at all having primary quills rounded at the extremities, is able to harvest success by North wind.

A middling-sized or a large head, slightly flattened, with a gracious curve, with a correlation circle white or yellow, or without that circle, shows a bird which will classify itself according to the condition of its feathers, and especially of that of the primary quills, which must be well rounded at the extremities.

IMPORTANT REMARKS

Do not keep more pigeons that you can observe and keep. Do not keep many but good ones.

To get a founder of a colony, apply to a famous but trusty breeder.

If you buy pigeons, buy by preference a couple of breeders, and mate the products observing the rules given in this book.

Avoid the old subjects and half-value.

Observe if the wings are in proportion to the weight, viz. if there is a correlation between the mass to be moved and the sails which have to lift it up and to carry it. Extend the wing and release it suddenly to make sure that the bird brings it back to the body with a brusk movement.

Look if the tail is well formed and if the bones of the abdomen are narrow.

If all these qualities are present in a bird, you can be sure of having a subject fit for fast flying. It must give the impression that it is held only by the tail and the top of the wings when you have it in your hands.

Thus : head strong and long, neck of medium length, strong wings, eye see picture 26-27-28.

When you add a new subject to your colony, give it a bath with lukewarm water.

Take information about its food from its former owner ; a change in the diet could be detrimental.

When you change the mixture according to the period, do it moderately, give small seeds in the morning and big ones at night.

Mix the seeds yourself, put them in open troughs, well aerated and in full light. Shake them often and before distributing, sift them.

Before buying the seeds make sure of their quality.

Put a handful of each sort somewhere where the temperature is good, on a small wet linen cloth.

If good, the seeds will germ in the proportion of 75 % in twice 24 hours.

Water must be clean.

Put every day fresh water into the water-fountain.

Keep the house clean ; the feaces are centers of propagation for microbes and parasites.

Allow no draft, give plenty of air and of sunlight.

Give the necessary condiments, specially greens every day during the resting time and the day following the return during the racing-season.

Give a lukewarm bath at least once a week by preference the day following the race.

REMARKS

Let us add for the benefit of the amateurs, who, with us will pursue their researches a few intersting remarks as to the training.

Respiration takes place about 16 to 36 times a minute.

The heart beats 120 or 150 times a minute.

The average temperature is 42° centigr.

If the amateur weights his subjects before and after each race and keeps a note book, the special thermometre about which I shall speak later, will indicate if your pigeon **is fit.**

It is up to you dear reader, to make all the required observation as well at the start as at the return.

This will help you greatly.

Keeping pigeons is no longer a matter of routine, without taking into consideration the enormous progress made by this sport.

How to Increase the Flying Speed of the Pigeon

I have made since 1900 numerous experiences, specially on the tail and on the wings, and therefore I invite the reader to keep up a booklet of personal remarks.

It is impossible to set fixed rules, the weight varying between 350 to 500 grs. and with the same subject differing with the seasons. You must also not compare the weight of a fat pigeon with that of a muscular one.

It is however certain that a relation exists between the wing surface and the weight.

It is thus possible to find the optimum charge and to encrease the speel of the pigeon which has too large a wings' surface, by reducing this surface, taking into account the forward resistance which diminishes with the elongation and then the diminishing of the length of the wings encreases this elongation.

The span of a good racing-pigeon is 69 to 70 cms. and experiences have been carried out to encrease the weight of pigeons being too light to a load of 8 to 9 kgs. per square meter.

By weight of the pigeon, we understand the weight taken during the good season, in the morning, before feeding, i. e. in the same condition as the birds are before the release in a race.

The surface of the wings normally extended, varies from 500 to 600 sq. cms.

By our pigeons, a load of 350 grs. for this surface gives only a low speed, the load should be 400 to 450 grs. in order to obtain better performances.

I have obtained very good results with pigeons which were always slow, by reducing the width of the wing, and even simply for a speed contest, by cutting 1 cm. on the width of the tail. For certain birds, the result was wonderful.

A second cm. can even sometimes be cut on the width of the tail.

These experiences cannot be carried too far, the fatigue becoming then too great even on a small distance, the motive power being limited, and decreasing with the duration of the effort.

Never cut the primary quills, this would decrease the span, but work on the secondary quill feathers, in order to make the wings narrower.

In order to know how much surface is to be taken away, in order to obtain the right proportion, act in the following way. Expend the wing, the back on a white sheet of paper, until the quills begin to separate.

Let the outline be drawn by an other person, and unit the extremities of the curve by a straight line.

On millimetred paper, the calculation is easy and rather accurate if you are patient.

Amateur, able to do do so wanting to buy a planimetre, will be able to determine the surface, using for this geometry, I mean calculating the total surface of the rectangles, triangles and trapezoids contained in the surface of the wings.

Certain amateurs cut out the wing traced on the paper, weight it and compare to the weight of a known surface of a same paper.

Multiply by 2, (there are 2 wings), if the birds is well balanced and you obtain the exact measurement of the wings' surface.

Weight the pigeon, when fasting, shutting it up in a basket where it cannot move and do not forget deducting the weight of the basket.

The ratio is obtained by dividing the square root of the surface by the cube root of the weight.

This ratio varies little and seems to be 3,5 as an average for the pigeon.

It remains only to calculate how much surface is to be suppressed from the wings, to be in the required proportion, which is 8 to 9 Kgs. per square metre.

This of course is only theoretical, and practically the motive power, the quality of the quills, the wind, the will of the bird have a great part in the success.

The experience is always a ticklish one, for a mistake in the calculation, once the wing is cut, is irremediable.

Observe carefully the drawning of the wing before cutting.

I advice you to be prudent and if some day one is able to adapt with certainty the unit loads, to the duration of the flight, my method of breeding will reach these results by the selected races it will produce.

How to Manage a Pigeon-House of " Speed Pigeons „

The amateur who likes to play the « speed game », and has only a short time to devote to it, must built small houses well fited, of the size stated in the second chapter.

In the room where these holes are placed, keep a place for the youngs of the year, which will have an outlet by an opening located beside all these little houses.

This compartment will be accessible to every subject.

In this compartment will be placed 2 or 3 unmated cocks, which will have no own hole, and which in order to go out, will have to cross the gangway of some couple ; this will make them very jealous.

In order to give a rest to these widowers, fix for each of them a separation board, fixed directly to the wall opposite the holes, so that at any moment all the subjects can see them.

Settled so, these little houses will have advantages on any other type.

Being fixed on the same wall they permit an easy inspection and give marvellous results for the return of all subjects after the race.

They are cheap, answer the hygienical requirements and each hole being reserved for a couple, the dangers of epidemics are reduced to a minimum.

Give them the preference. They are suitable for :

Any sort of game, but specially for the speed races.

Are considered as speed contests, every races shorter than 300 Kms.

These little houses will be dismounted as well as all the rest of the material before and after the racing season.

During the resting period, all the subjects will stay in the appartment.

Only the opening reserved to the youngs of the year will be left free ; all the other being obstructed until the couples are mating again.

Mate at the herebefore given date.

The food for these subjects will be the mixture given in the second book.

If a bird must feed a young, encrease the ration and observe the birds, in order to avoid that they get fat, but keep hard muscles and be healthy.

Give a bath with lukewarm water, at least once a week, and during the racing season, the day following the return from a race.

All the subjects of these small pigeon-houses, which have to travel will partake in the daily morning and evening flight, though never tyring them.

During these flights, clean thoroughly the house, and each hole.

Fresh water will be given in a clean trough (1 per couple). The ration will be given in another trough, (1 per couple).

The gravel trough will contain all the required condiments, and be placed in the centre of the appartment for the whole colony.

The widowers in the great house, kept to excite the jealousy of the racers, will have the same feathers as these ones and will be locked from time to time in the small holes, during the exercises, i. e. during the absence of the other birds.

They will remain there until the bird puts it out. Watch the birds and do not let them fight, for the widowers once installed are often very vicious and could put out one of your birds' eye or wound it so that it could not partake to the races anymore.

These widowers having similar feathers as the carrier pigeons, will be placed for the speed contest, according to their shades, with the hens of the carriers, and these ones instead of turning them away will caress them.

This will render the carrier terribly jealous and returning to the house, this one will fly straight home.

Any game can be practised in those small houses. You will refer to the instructions given in the dispositions of a bird for such and such a distance. With some initiative, you will observe that the combinaisons are endless in this system of game.

————

Supplement

This is the end of the last book of the « Explained selection and the leading of the Pigeon-houses during the Four Seasons » or simply « The Four Seasons » real course about racing-pigeons, by Mr. J. Heuskin.

Its success is certain, while it explains things which every-body must learn, who thought to know them.

Please listen to this little story.

One of my friends had hired a gardener. The first thing this man did, was to turn the mint border.

Then, he started at the rubarbs, but my friend gave him warning. He was not gardener, but simply an unemployed mason.

Of course my friend was furious. Why did he pass himself as a gardener ? Why not after all ? They are thousands of people calling themselves breeders, who do not know the art of sorting pigeons, more than this gardener knew his job.

It is a fact that new ideas and methods come up in all provinces, but in pigeons breeding these ideas were not amalgamated, while since about 10 years, everybody asks for a true, a good method, which remained however undiscoverable.

However I had in my bookselling department all the serious authors, such as : André, Gigot, Henin, Ratio, Van

Linden, Violette, Wittouck, etc., to teach the pigeons's friends, who feel more and more, that if it was well to read the books, the practice, which was the sole rule of the ancient breeder being not sufficient.

Then, I found a thorough observer, who had a sufficient practise based on an experience of 40 years to be able to recognize the truth of his theory.

He communicated it to me and sometimes his conceptions are still open to discussion. However though not being a beginner, I was very pleased to talk to this man, and thought that every amateur would experience the same pleasure.

I believe, I was right if I considered the felicitations and testimonials I got from the first breeders. The author has granted me after the first volume has been issued, the sole selling right for Belgium and foreign countries.

I could have presented this book in a more elaborate way, but I would perhaps have been less clear.

The amateur will read and read again the book, and will find in it every information he may need.

Do not find difficulties everywhere.

You will find in this book, daring methods such as the forced bath, which will be hotly discussed by certain competences. Do not forget that if you follow them exactly, you will run no risk, all what is stated in this book having been controled.

Without mentionning the classification and other details which, as has been the case with me, will have interested you to the highest degree ; the evolution of the mysterious point, is amongst others a proof, that observation leads to any discovery.

Adopt one or another system of game, follow strictly the method given, and you will soon notice its advantages.

Reckon only on yourself.

Marvellous pigeons are scarce.

You will only get them through cultivating your colony and the improving elements will be chosen by you, with a thorough knowlegde of the matter, knowing now all the qualities required to make a good bird.

Surely, the « Four Seasons » readers are keen amateurs and I expect a series of triumphs, to be the result of the application of these methods.

If we go back to the first years of our sport, except a few exceptions, whose name are still well known, who however have left no method, the results and success were dependent on personal experience alone, i. e. empiricism and this from one generation to the other, always beginning anew, without any practical knowlegde, without taking advantage of the disappointments and the struggles of the ancient breeders.

To day, observe the mistakes of others, study their inventions, incorporate their experience cost next to nothing. This leads to quite interesting researches.

The reader will have found in this book directions and suggestions which can be used in his pigeon-house, and he will reach that happy state, for wich he spent long hours, and which would have cost him much money and time, if he had been left to his own device.

It has been objected, that racing would become very difficult, but this is progress !

The main thing is to begin with sound principles and to reserve the same enthusiasm tô the success of the others as to its own, and to reserve his whole energy to the greater things in the future ; for the real breeder likes always

competition with a superior competitor, and the best amateur is this one who likes to race his fit subjects with the
best pigeons of his antagonist.

I shall go on for the sake of the sport, as in the past,
to gather all sound principles leading to success.

Eug. BUCHMANN.

Publisher,

Manager of the Manufacture of Articles for Pigeons,

140, Rue Renory — Kinkempois (Belgium).

Bookselling departement.   JANUARY 1930.

CONTENTS

www.ingramcontent.com/pod-product-compliance
Lightning Source LLC
Chambersburg PA
CBHW030633270326
41929CB00007B/61